Michael Stefan Rill

Three-Dimensional Photonic Metamaterials

Michael Stefan Rill

Three-Dimensional Photonic Metamaterials
by Direct Laser Writing and Advanced Metallization Techniques

Südwestdeutscher Verlag für Hochschulschriften

Imprint
Any brand names and product names mentioned in this book are subject to trademark, brand or patent protection and are trademarks or registered trademarks of their respective holders. The use of brand names, product names, common names, trade names, product descriptions etc. even without a particular marking in this work is in no way to be construed to mean that such names may be regarded as unrestricted in respect of trademark and brand protection legislation and could thus be used by anyone.

Publisher:
Südwestdeutscher Verlag für Hochschulschriften
is a trademark of
Dodo Books Indian Ocean Ltd., member of the OmniScriptum S.R.L Publishing group
str. A.Russo 15, of. 61, Chisinau-2068, Republic of Moldova Europe
Printed at: see last page
ISBN: 978-3-8381-1888-8

Zugl. / Approved by: Karlsruhe, Karlsruhe Institute of Technology (KIT), Diss., 2010

Copyright © Michael Stefan Rill
Copyright © 2010 Dodo Books Indian Ocean Ltd., member of the OmniScriptum S.R.L Publishing group

Contents

Abbreviations   iii

1. Introduction   1

2. Electrodynamics of Magneto-Dielectric Materials   5
    2.1. Fundamentals of Effective Media 6
    2.2. Magnetic Response of Effective Media 9
    2.3. Properties of Isotropic Magnetic Materials 10
    2.4. Design of Magnetic Building Blocks 13
    2.5. Purely Magnetic Metamaterials 18
    2.6. Negative-Index Metamaterials 19
       2.6.1. Passive Medium Conditions 20
       2.6.2. Drude Model and Diluted Metals 22
    2.7. Design of Negative-Index Metamaterials 25
    2.8. Bi-Anisotropic Metamaterials 26
       2.8.1. "Pure" Bi-Anisotropy 26
       2.8.2. Chirality 29

3. Three-Dimensional Metamaterials for Photonics   33
    3.1. Bulk or Non-Bulk: That's the Question. 34
    3.2. Layer-By-Layer Approaches Towards Three-Dimensional Metamaterials 35
       3.2.1. Single-Step Structuring 36
       3.2.2. Planarization 36
    3.3. Inherent Three-Dimensional Fabrication 36
       3.3.1. Three-Dimensional Polymer Templates by Direct Laser Writing 38
       3.3.2. Protection of Polymers by Highly Stable Oxide Ceramics 40
       3.3.3. Metallization of Three-Dimensional Polymer Templates 42
    3.4. Analysis of Photonic Metamaterials 46
       3.4.1. Calculation of Electromagnetic Fields and Optical Spectra 47
       3.4.2. Measurement of Optical Spectra 49

4. Bi-Anisotropic Three-Dimensional Metamaterials   53
    4.1. Analytic Model of a Bi-Anisotropic Split-Ring-Resonator Array 54
    4.2. Fabrication and Optical Characterization 56
    4.3. From Isolated Split-Ring Resonators to Corrugated Metal Surfaces 60
    4.4. Negative-Index Bi-Anisotropic Metamaterial 63
    4.5. Interim Result 68

5. Towards Bulk Photonic Metamaterials   69

6. Conclusions and Outlook   **73**

A. Background and Details   **77**
   A.1. Wood Anomaly 77
   A.2. Reciprocity in Optics 78
   A.3. Lorentz Oscillator Model 79
   A.4. Mathematical Proof that (2.39) Solves (2.38) 80
   A.5. Fresnel Equations of Purely Bi-Anisotropic Media 82
   A.6. Numerical Time-Domain Calculations 85

Bibliography   **89**

Acknowledgments   **101**

Abbreviations

ALD Atomic layer deposition (ALD) is a thin film coating technique based on sequential, self-limiting surface reactions. It enables the control of film thicknesses on the atomic scale resulting in conformal films even on high-aspect ratio structures.

CVD "Chemical vapor deposition" (CVD) denominates a process used to deposit high-purity solid materials. In a typical CVD process, the substrate is exposed to one or more volatile precursors which react and / or decompose on the substrate surface. In many cases, by-products are also produced which are either removed by a carrier gas or by evacuation of the reaction chamber.

DLW Direct laser writing (DLW) is a lithography technique for fabricating structures in a photoresist on the microscale, without requiring complex optical systems or photomasks. DLW relies on a multi-photon absorption process changing the solubility of the resist in the high-intensity focal spot of a laser beam. Scanning the sample (or the laser) relative to the laser focus results in connected three-dimensional patterns after development.

EBL Electron-beam lithography (EBL) utilizes an electron beam to structure a surface covered with a resist. Selectively removing either exposed or non-exposed regions of the resist yields high-quality nanostructures. EBL is widely used as a maskless tool for low-volume production of semiconductor components as well as research and development.

FIB Focused-ion beam (FIB) systems operate in a similar fashion to scanning-electron microscopes (SEM) but use a focused beam of ions (usually gallium). The beam of atoms can be either operated at low currents for imaging purposes or at high beam currents for local sputtering and milling. While the ion beam scans the sample surface, the signal from the sputtered ions and / or the secondary electrons are collected to form an image.

FOM The Figure of Merit (FOM) is, in general, a measure for the performance of a system. Relating to metamaterials, the FOM is defined as the negative ratio of real and imaginary parts of the refractive index n, i.e., FOM$=-\text{Re}(n)/\text{Im}(n)$. High *positive* values are obtained if the respective metamaterial structure features high *negative* values for $\text{Re}(n)$ while keeping the absorption (determined by $\text{Im}(n)$) relatively low.

FTIR Fourier-transform infrared spectrometers (FTIRs) use an interferometer (e.g., Michelson interferometer) at which the emitted light from a thermal source is split into two partial beams. Both beams are reflected at mirrors, whereas one of those mirrors is continuously wobbling. After both beams have interfered, they pass the sample and are finally detected. The intensity is measured *versus* the optical path difference Δx caused by the moving mirror. Subsequently, the obtained interferogram

	is Fourier transformed to the frequency domain to obtain either the transmittance or the reflectance spectrum.
IR	Electromagnetic radiation with a wavelength between 0.7 µm (converting to a frequency of $\nu = 430\,\text{THz}$) and 300 µm ($\Rightarrow \nu = 1\,\text{THz}$) is assigned to the infrared spectral range (IR).
ITO	Indium tin oxide (ITO) is a solid solution of indium$^{(III)}$ oxide (In_2O_3) and tin$^{(IV)}$ oxide (SnO_2), typically 90% In_2O_3, and 10% SnO_2 by weight. The material is transparent and colorless in thin layers while exhibiting electrical d.c. conductivity.
PLD	"Pulsed layer deposition" (PLD) denominates a coating technique similar to atomic layer deposition (ALD), but delivering much thicker layers (typically some nanometers in thickness) during each reaction cycle.
SRR	The split-ring resonator (SRR) is an electro-optical building block exhibiting strong magnetic resonances which can be used to obtain a permeability unequal to one.
VIS	Electromagnetic radiation being visible to the human eye is assigned to the visible spectral range (VIS). A typical human eye will respond to wavelengths from about 380 nm (converting to a frequency of $\nu = 790\,\text{THz}$) to 750 nm ($\Rightarrow \nu = 400\,\text{THz}$).

1. Introduction

> I exhorted all my hearers to divest themselves of prejudice and to become believers in the Third Dimension ...
>
> *(Edwin A. Abbott, 1884,*
> *from: "Flatland. A Romance of Many Dimensions")*

Since time immemorial, mankind makes use of artificial optical devices. In fact, a well-preserved mirror was found next to pharaoh Senusret's pyramid that is around 4000 years old [1]. Other archaeological excavations revealed that plano-convex and sphere-shaped lenses were already known to the ancient Romans. The targeted manipulation of light paths by using natural substances has, apparently, a very long history. However, in those days, it was by far not understood what kind of physical and chemical mechanisms give rise to the respective optical effects. Thus, optical science had mainly based on observational research, experiences passed on from generation to generation, as well as trial and error.

With the beginning of the Age of Enlightment, many scientific disciplines received a substantial boost from new pioneering ideas. Besides that, also nearly forgotten concepts like "the atom" experienced a comeback. The earliest references on atoms being the basic building blocks of matter trace back to the old Indian and Greek philosophy (600 and 450 B.C., respectively). But merely since the middle of the 19th century, scientific theories and experiments have reached a level of sophistication to really prove that matter consists of smaller constituents whose interaction, relative spatial position, and kinetics affect all of its intrinsic characteristics.

Parallel to the achievements in atomic physics, Michael Faraday and James C. Maxwell triggered the era of modern optics by understanding light as an electromagnetic wave. In 1878, Henrick A. Lorentz had the idea to consider atoms as oscillating dipoles, thus, successfully linking the observed material response to the wave-like nature of light. The experimental verification of Maxwell's electromagnetic theory and the oscillator model attributes to Heinrich Hertz who succeeded in generating and detecting radio waves in 1887.

Ever since, these fundamental discoveries inspired many scientists to successively improve and extend the physical understanding of light-matter interactions. The acquired knowledge and the progressing experimental developments finally enabled the targeted manipulation of material properties. Prominent examples of designed material systems that emerged during the last decades include doped semiconductors [2,3], gain media for laser applications [4], and photonic crystals [5–8]. Importantly, the manipulations of optical attributes have primarily acted on the variation of the dielectric response (also known as the permittivity ε) to electromagnetic radiation. The corresponding magnetic response (also known as the permeability μ) was rather neglected, thus, fading out many possible effects from the outset.

In the late 1960s, Victor G. Veselago acknowledged this problem as he derived novel, exotic properties for hypothetic materials whose permittivity and permeability are simultaneously negative, thus, leading to a negative refractive index n [9]. A negative permittivity was already known for metals below their plasma frequency. However, an analog behavior for the permeability requires strong coupling to the magnetic component of light. For natural materials, this had never been observed in the infrared or visible spectral range. Hence, the concept initially fell into oblivion. Indeed, it took 30 years until John Pendry re-discovered Veselago's publication. Pendry adapted known concepts from high-frequency technology and proposed a way to provide resonant magnetic coupling even in the optical regime [10]. For this purpose, he utilized periodically-arranged metallic LCR circuits which he called *split rings*. These "U"-shaped wires [11] are meant to serve as unit cells of a composite—like atoms and molecules in a solid. If they are much smaller than the wavelength of the incident light, the effective medium formalism of optical materials can be applied. Hence, this concept opened the door for artificial composite media whose elementary building blocks can be customized at will. These so-called *metamaterials* enable optical properties not provided by natural materials.

Fortunately, the timing of Pendry's proposal was absolutely perfect since micro- and nanofabrication techniques were readily available. Thus, the experimental realization of metamaterials was within reach. In 2000, David R. Smith an co-workers demonstrated negative-index composite structures at microwave frequencies of about 10 GHz for the first time [12–14]. Since then, metamaterials have attracted a lot of interest which was mainly driven by the fascinating visions of perfect lenses [15, 16], the inverse Doppler effect [17], or the reversed Čerenkov radiation [18, 19]. Beyond Veselago's predicted optical effects, also new approaches based on transformation optics have caused a lot of excitement in the context of optical cloaking [20–23].

Many of the observed phenomena are considered to be only of academic interest. However, there has been always a quest of finding applications which might also have an impact on the society in general. Therefore, it is certainly important to revert to metamaterials working in the optical / telecommunication regime. Light in this spectral region has a wavelength of around 1 µm. To fulfill the effective medium condition, the unit cells of the composites must be much smaller than that. Thus, how can we actually come up to such tiny feature sizes? The vast majority of metamaterials for the optical spectral range [24–27] has been fabricated by electron-beam lithography and physical vapor deposition of metal films, both of which are well-established two-dimensional (2D) nanotechnologies. Resulting structures consist of a single layer of planar unit cells with only several nanometers in thickness. In this case, the magnetic response changes the phase of a passing electromagnetic wave only marginally. The observed properties are rather dominated by surface effects. However, to considerably enhance the contribution of propagation effects, bulk metamaterial structures are required. Therefore, several groups of the photonic community "became believers of the Third Dimension"[1] and began to work on the experimental implementation. Proposed solutions mostly utilize 2D lithography combined with planarization [28, 29] or one-step structuring techniques [30, 31]. For practical reasons, however, it would be preferable to replace the 2D processes by their inherently 3D analogues [32, 33]. A corresponding approach is given in the course of this Thesis.

---

[1]Freely adapted from E. A. Abbott

Outline

In chapter 2, the electromagnetic theory of magneto-dielectric materials is presented. We will start with general assumptions on the material response to electromagnetic radiation and focus on the magnetic response. Indeed, metamaterials are an excellent system to investigate the derived effects. Next, we will further extend our discussions to sufficient and necessary conditions for obtaining a negative refractive index. Finally, the properties of chiral and purely bi-anisotropic materials are considered. The emerging results will be of great importance for the characterization of the fabricated structures.

Chapter 3 deals with the fabrication of photonic metamaterials. The demand for bulk composites brought up several concepts which will be discussed and evaluated. Additionally, our proposal of fabricating 3D structures by using inherently 3D processes will be presented, i.e., direct laser writing combined with advanced coating techniques.

In chapter 4, we use our fabrication method to realize an array of 3D split rings and related descendants. As these structures are not symmetric along the propagation direction of the incident light, they must be treated as bi-anisotropic media. To relate the polarization (magnetization) of a reciprocal bi-anisotropic structure to the magnetic (electric) field, additional cross-coupling parameters are required.

A first attempt to realize bulk metamaterials is presented in chapter 5. Although the feature sizes are not suitable to obtain a negative refractive index yet, we show anyhow that our approach opens new possibilities for further investigations. Finally, in the last chapter, the aspects of the Thesis are summarized and an outlook on further experimental improvements and research activities is given.

2. Electrodynamics of Magneto-Dielectric Materials

In the course of this Thesis, we study electromagnetic waves in space occupied by matter [1,34,35]. Like other macroscopic theories, classical electrodynamics is concerned with physical quantities averaged over small volumes. Hence, any microscopic variation of the quantities caused by the unit cells' (molecular) fine structure is neglected. For simplicity, all following discussions are related to matter consisting of periodic unit cells (e.g., crystalline solids), where the ratio of the lattice constant a and the vacuum wavelength of the incident light beam λ is a measure of how to treat the underlying system.

We will call the illuminated medium "effective", if the condition

$$a \ll \lambda \tag{2.1}$$

holds. For natural substances and the visible spectral range (VIS), this is the common case since the ratio of λ/a is in the order of 1000. When light is propagating through such an effective medium, the microscopic electric and magnetic fields of the molecules can be homogenized[1]. Thus, the material's response to the incident light wave can be assigned to effective parameters like, e.g., the refractive index $n := c_0/c_\mathrm{m}$. Here, c_0 and c_m denote the phase velocities of light in vacuum and inside the medium, respectively. We will also become acquainted with other material parameters which explicitly distinguish between the light wave's electric and magnetic fields.

If (2.1) is *not* fulfilled, the light wave is able to resolve the material's atomic structure. For example, the crystallographic structure of natural solid crystals can be resolved by X-rays whose vacuum wavelength is much smaller than the lattice periodicity ($\lambda_\text{X-ray} \ll a_\text{solid} \approx 0.1\,\text{nm}$). Hence, the effective medium description must be replaced by a band structure approach if diffraction appears.

But how much larger than a must λ be to assure that the effective medium theory provides reliable results? Actually, this question has caused many controversial discussions: Ref. [36] shows, e.g., that the effective medium theory even holds true for periodic materials with $\lambda/a \approx 1$. Ref. [37] discusses effective material parameters for a unit cell size of $\lambda/a > 10$, whereas Refs. [38, 39] claim that effective optical parameters cannot be applied to structures where $\lambda/a < 100$. Indeed, in situations where $\lambda/a < 100$, we have to deal with significant phase retardation across each unit cell.

---

[1] This is the so-called "first homogenization". Later, in the context of composite materials, we will also introduce a "second homogenization" where the light fields can be averaged over *mesoscopic* unit cells of a *macroscopic* structure.

Nevertheless, there must exist a transitional regime between a "perfect" effective medium and a band structure description of periodic structures, where the definitions of effective material parameters are approximate but still helpful in the interpretation of the medium's scattering properties. As long as we keep away from Wood anomalies (further details in section A.1), the incident wave will not be diffracted. Thus, it is justified to consider such materials as "quasi-effective". The according condition is much weaker than (2.1) and solely determined by the first diffraction order

$$a < \frac{\lambda}{n_{\text{bg}}}, \tag{2.2}$$

where n_{bg} is the refractive index of the background medium. On this condition, the quasi-effective materials interact with electromagnetic radiation in a similar manner as effective materials would do. In other words: By looking at the incident and outgoing light waves, we determine the optical properties of a "black-boxed" material which must not necessarily fulfill the effective medium condition (2.1).

2.1. Fundamentals of Effective Media

We consider monochromatic electromagnetic waves impinging from vacuum onto a slab of a dispersive effective medium. In this case, all components of the electric and magnetic fields can be expressed as harmonic functions of time with the same frequency $\omega = 2\pi\nu = 2\pi c_0/\lambda$, i.e.,

$$\vec{E}(\vec{r}, t) = \text{Re}\left(\vec{E}(\vec{r})\, e^{-i\omega t}\right), \tag{2.3}$$
$$\vec{H}(\vec{r}, t) = \text{Re}\left(\vec{H}(\vec{r})\, e^{-i\omega t}\right). \tag{2.4}$$

$\vec{E}(\vec{r}, t)$ and $\vec{H}(\vec{r}, t)$ correspond to vector amplitudes of the electric and magnetic field, respectively. When the incident electromagnetic wave penetrates transparent matter, electric and magnetic dipoles are excited which re-emit electromagnetic waves just like an antenna. The re-emitted radiation excites other dipoles so that repeating this mechanism leads to propagation of light inside the medium. Here, the incident light wave drives non-resonant oscillations of the atoms at its own frequency. The atomic oscillations follow those of the driving wave but with a certain phase lag which accumulates through the medium and, hence, retards the propagation of the wave front. Clearly, these interactions modify the velocity of light and, thus, determine the optical properties.

To describe the internal fields inside an effective medium, the electric displacement $\vec{D}(\vec{r}, t)$, the magnetic induction $\vec{B}(\vec{r}, t)$, the polarization $\vec{P}(\vec{r}, t)$ as well as the magnetization $\vec{M}(\vec{r}, t)$ have to be introduced. These field vectors are related to the incident light fields $\vec{E}(\vec{r}, t)$ and $\vec{H}(\vec{r}, t)$ via

$$\vec{D}(\vec{r}, t) = \varepsilon_0 \vec{E} + \vec{P}(\vec{E}, \vec{H}), \tag{2.5}$$
$$\vec{B}(\vec{r}, t) = \mu_0 \left(\vec{H} + \vec{M}(\vec{E}, \vec{H})\right), \tag{2.6}$$

where ε_0 and μ_0 denote the vacuum permittivity and vacuum permeability, respectively. For reasons of readability, we omit the explicit space and time dependence of the fields on the right-hand sides of (2.5) and (2.6). The polarization (magnetization) is defined as the sum over all individual electric (magnetic) dipole moments $\vec{p}$ ($\vec{m}$) times the number density of the dipoles. In contrast to most textbooks on optics, we assume $\vec{P}$ and $\vec{M}$ to depend on both the electric and the magnetic field, since we will also encounter bi-anisotropic and bi-isotropic (chiral) media. For details, we refer to section 2.8.

The incident fields ($\vec{E}$ and $\vec{H}$) can be considered as inputs which induce a motion of charges inside the medium. The resulting collective movement in dispersive media gives rise to a polarization (and a magnetization, respectively) that is a superposition of the effects of $\vec{E}(t')$ and $\vec{H}(t')$ for all times $t' \leq t$. As we refrain from using high light intensities, the response on the incident fields is assumed to be linear and local. Thus, the polarization and magnetization are given by

$$\vec{P}(\vec{r},t) = \varepsilon_0 \int_{-\infty}^{t} \underline{\chi}_e(\vec{r},t-t')\,\vec{E}(\vec{r},t')\,\mathrm{d}t' + \frac{1}{c_0}\int_{-\infty}^{t} \underline{\xi}(\vec{r},t-t')\,\vec{H}(\vec{r},t')\,\mathrm{d}t' \,, \tag{2.7}$$

$$\vec{M}(\vec{r},t) = \int_{-\infty}^{t} \underline{\chi}_m(\vec{r},t-t')\,\vec{H}(\vec{r},t')\,\mathrm{d}t' + \frac{1}{\mu_0 c_0}\int_{-\infty}^{t} \underline{\zeta}(\vec{r},t-t')\,\vec{E}(\vec{r},t')\,\mathrm{d}t' \,. \tag{2.8}$$

If the material is considered to be homogeneous, the susceptibilities simplify to $\underline{\chi}_e(\vec{r},t) = \underline{\chi}_e(t)$ and $\underline{\chi}_m(\vec{r},t) = \underline{\chi}_m(t)$. By using the convolution theorem, we can rewrite (2.7)–(2.8) as

$$\vec{P}(\vec{r},\omega) = \varepsilon_0 \underline{\chi}_e(\omega)\,\vec{E}(\vec{r},\omega) + \frac{1}{c_0}\underline{\xi}(\omega)\,\vec{H}(\vec{r},\omega) \,, \tag{2.9}$$

$$\vec{M}(\vec{r},\omega) = \underline{\chi}_m(\omega)\,\vec{H}(\vec{r},\omega) + \frac{1}{\mu_0 c_0}\underline{\zeta}(\omega)\,\vec{E}(\vec{r},\omega) \,, \tag{2.10}$$

where $\underline{\chi}_e$ ($\underline{\chi}_m$) denotes the electric (magnetic) susceptibility tensor. $\underline{\chi}_e$ ($\underline{\chi}_m$) describes the electric (magnetic) response of the material to the incident electric (magnetic) fields. In many situations, it is useful to combine the pure material responses and the incident fields. The permittivity tensor $\underline{\varepsilon} = \mathbb{1} + \underline{\chi}_e$ is defined as the total electric response. Accordingly, the permeability tensor $\underline{\mu} = \mathbb{1} + \underline{\chi}_m$ defines the total magnetic response. Electric dipoles can be, in general, also excited by the magnetic field resulting in a non-zero cross-term tensor $\underline{\xi}$. Likewise, magnetic dipoles can be excited by electric fields being analogously specified by $\underline{\zeta}$.

Notably, all mentioned material parameter tensors generally have complex entries as they should also be applicable for lossy media. As long as no static fields are present, the material parameter tensors are directly related to each other via reciprocity [40]. Concretely, this means that $\underline{\varepsilon} = \underline{\varepsilon}^t$, $\underline{\mu} = \underline{\mu}^t$, and $\underline{\zeta} = -\underline{\xi}^t$ [41], where the superscript t denotes a transposed tensor (for details q.v. section A.2). Indeed, for all following discussions, these relations will be always fulfilled. Depending on the mathematical properties of $\underline{\varepsilon}$, $\underline{\mu}$, $\underline{\xi}$, and $\underline{\zeta}$, all media can be classified by their isotropic, anisotropic, bi-isotropic, or bi-anisotropic optical behavior. A brief overview is given in Tab. 2.1.

By using the Fourier-transformed definition of (2.5)–(2.6) as well as (2.9)–(2.10), we finally obtain

$$\vec{D}(\vec{r},\omega) = \varepsilon_0 \underline{\varepsilon}(\omega)\,\vec{E}(\vec{r},\omega) + \frac{1}{c_0}\underline{\xi}(\omega)\,\vec{H}(\vec{r},\omega) \,, \tag{2.11}$$

$$\vec{B}(\vec{r},\omega) = \mu_0 \underline{\mu}(\omega)\,\vec{H}(\vec{r},\omega) - \frac{1}{c_0}\underline{\xi}^t(\omega)\,\vec{E}(\vec{r},\omega) \,. \tag{2.12}$$

Table 2.1.: Classes of media categorized by their respective optical parameters. Isotropic media are uniform (homogeneous) in all directions, whereas anisotropy is the property of being directionally dependent. Bi-isotropy as well as bi-anisotropy are related to the existence of magneto-electric coupling effects. Detailed information can be found in section 2.8.

isotropic media	anisotropic media
$\varepsilon, \mu \in \mathbb{C}$	$\underline{\varepsilon}, \underline{\mu}$ are tensors
$\xi, \zeta = 0$	$\underline{\xi}, \underline{\zeta} = \mathbb{O}$
bi-isotropic (chiral) media	**bi-anisotropic media**
$\varepsilon, \mu \in \mathbb{C}$	$\underline{\varepsilon}, \underline{\mu}$ are tensors
$\xi, \zeta \in \mathbb{C}$	$\underline{\xi}, \underline{\zeta}$ are tensors

These are the general constitutive material equations for reciprocal effective media with local response.

So far, we have only looked at the material response to incident fields, yet being unconcerned about the propagation of electromagnetic waves itself. By using the time-harmonic fields (2.3) and (2.4), the general Maxwell equations link the electric fields to the magnetic fields in a succinct manner given by

$$\nabla \cdot \vec{D}(\vec{r},\omega) = \varrho(\vec{r},\omega) \,, \quad (2.13)$$
$$\nabla \cdot \vec{B}(\vec{r},\omega) = 0 \,, \quad (2.14)$$
$$\nabla \times \vec{E}(\vec{r},\omega) = -i\omega \vec{B}(\vec{r},\omega) \,, \quad (2.15)$$
$$\nabla \times \vec{H}(\vec{r},\omega) = \vec{j}(\vec{r}) + i\omega \vec{D}(\vec{r},\omega) \,, \quad (2.16)$$

where $\varrho(\vec{r},\omega)$ and $\vec{j}(\vec{r},\omega)$ are the charge and current density, respectively. With (2.13)–(2.16), we have relations at hand allowing to derive and solve the differential wave equation which again entirely determines the propagation of light. If the fields are influenced by the general constitutive equations (2.11)–(2.12), a solution for the wave equation cannot be found. Only in case of the problem being considerably eased to, e.g., a *uniaxial* bi-anisotropic configuration, the wave equation can be solved by means of plane wave expansion [41–43] or vector transmission-line theory [44]—albeit the derivations are quite lengthy.

By default, in textbooks on optics, the wave equation is derived for isotropic dielectrics. This is a fairly simple case since $\varrho(\vec{r},\omega)$ and $\vec{j}(\vec{r},\omega)$ are strictly zero, the material parameters in (2.11)–(2.12) are real scalars, and the cross-term parameters ξ and $\zeta = \xi^t$ vanish. Thus, by arranging the curl of (2.15), the differential wave equation of the electric field results in

$$\nabla^2 \cdot \vec{E}(\vec{r},\omega) + \frac{\omega^2}{c_0}\mu(\omega)\varepsilon(\omega)\, \vec{E}(\vec{r},\omega) = 0 \,. \quad (2.17)$$

An alternative ansatz which uses the curl of (2.16) yields an analog differential equation for the magnetic field. Possible solutions of (2.17) are transverse electromagnetic plane waves (TEM) which are given by $\vec{E}(\vec{r}) = \vec{E}_0\, e^{i\vec{k}\vec{r}}$.

2.2. Magnetic Response of Effective Media

We have assumed that, in principle, light waves may couple to both electric and magnetic dipoles while propagating through an arbitrary medium. Regardless of this fact, in most textbooks on wave-optics the permeability $\underline{\mu}$ is set to unity. Therefore, almost all discussions are devoted to the variety of optical phenomena for this special case where only the permittivity tensor $\underline{\varepsilon}$ is considered. This assumption is based on the observation that for natural isotropic substances under standard environmental conditions, χ_m varies from zero at most in the order of 10^{-4} [2,3][2]. This fact was also comprehended theoretically by Landau and Lifshitz [35] who claimed that $\underline{\mu}(\omega)$ progressively loses its physical meaning of a response function as the frequency of the electromagnetic waves is increased. This is related to the properties of the total induced magnetic dipole moment $\vec{m}$ of a macroscopic body.

In the static case, the total magnetic dipole moment per unit volume corresponds to the magnetization $\vec{M} = \vec{B}/\mu_0 - \vec{H}$. This admits of introducing the permeability as a well-defined response coefficient—as discussed in the former section. If time-dependent fields are present, $\vec{m}$ is not only determined by the magnetization $\vec{M}$ but also by the time-dependent polarization $\vec{P}$. The induced current density consists now of *two* terms, i.e.,

$$\vec{j} \sim \underbrace{\vec{\nabla} \times \vec{M}}_{(I)} + \underbrace{(\partial \vec{P}/\partial t)}_{(II)} \,, \tag{2.18}$$

and so does the induced magnetic dipole moment

$$\vec{m} \sim \int \vec{r} \times \vec{j} \, \mathrm{d}V = \int \vec{M} \, \mathrm{d}V + \int \left(\vec{r} \times \frac{\partial \vec{P}}{\partial t} \right) \mathrm{d}V \,.$$

As a consequence, the association of $\vec{M}$ with the magnetic dipole moment of a unit volume depends on the possibility to neglect the contribution of the time-dependent polarization. If it is possible to separate the magnetic current (I) from the total induced current in (2.18), $\underline{\mu}(\omega)$ will retain its traditional physical meaning.

For a small macroscopic body consisting of a natural material (without resonant dispersion), Landau and Lifshitz compared both contributions (I) and (II) to the current density in a monochromatic time-dependent magnetic field. "Small" means that the body's characteristic length is much smaller than the incident wavelength ($l \ll \lambda$) but still macroscopic ($l \gg a$) with respect to the lattice constant a. In this case, the magnetic susceptibility of a homogeneous isotropic diamagnet is given by $\chi_m \sim a^2/\lambda^2$ and the electric susceptibility by $\chi_e \approx 1$. The deduced approximation [35, 47] reads

$$\left| \frac{\vec{\nabla} \times \vec{M}}{\partial \vec{P}/\partial t} \right| \sim \left(\frac{\lambda}{l} \right)^2 \frac{\chi_m}{\chi_e} \sim \left(\frac{a}{l} \right)^2 \ll 1 \,, \tag{2.19}$$

where the last inequality stems from the requirement of a macroscopic body. Apparently, the contribution of the polarization term (II) seems to be substantial.

---
[2]Notable exceptions in the microwave regime are found for rare earth compounds like $\mathrm{La_{1/3}Ca_{1/3}MnO_3}$ [45] and $\mathrm{LaMnO_3}$ (doped with strontium and iron, cobalt, or nickel) [46].

However, (2.19) is certainly not fulfilled if χ_m becomes significantly larger than estimated for natural diamagnets, e.g, for large-permittivity media [47] (i.e., $\chi_e \gg 1$). Indeed, in the following context, we will present some concrete examples for materials which have an intentionally strong magnetic response. Hence, attaining $\underline{\mu} \neq \mathbb{1}$ or even negative tensor components is not a conceptual problem at all, and identifying ways to realize materials showing such a strong magnetic response might revolutionize the field of optics and photonics.

Regrettably, there are no appropriate tools at hand to immensely enhance the magnetic response of atoms or molecules. If we want to tailor a material's electromagnetic property, we will have to build the functional "atoms" on our own. The resulting composites, the so-called "metamaterials"[3], would not only inherit the natural substances' characteristics but also introduce new effects stemming from their intrinsic geometry. In analogy to the former section, we will describe the light-matter interaction in terms of *mesoscopic* effective material parameters, as if the composites would fulfill all effective-medium conditions. Note that this relates to a second homogenization of the incident light field. Certainly, we must always remind ourselves that the unit cells themselves consist of non-magnetic media.

Optical parameters of metamaterials like the refractive index n, the impedance Z, and related quantities like the transmittance and the reflectance will surely change as the permeability varies. Hence, we should review some discussions from optical textbooks—usually concerned with non-magnetic dielectrics—and re-calculate fundamental relations for magnetic media. In the first instance, we will restrict our considerations to isotropic matter although the results will not be directly adaptive to the general bi-anisotropic case. This will facilitate to keep track of new appearing effects and will give us a coarse idea of potential concepts emerging. Nonetheless, some aspects of bi-anisotropic materials being important for the later discussions will be resumed in section 2.8.

2.3. Properties of Isotropic Magnetic Materials

As mentioned, for isotropic magnetic media, the material parameters are complex scalars. Moreover, both cross-term parameters ξ and ζ vanish and need not be taken into account (compare Tab. 2.1). These assumptions reduce the complexity of the general constitutive relations (2.11)–(2.12) to

$$\vec{D} = \varepsilon_0 \varepsilon(\omega) \vec{E} ,$$
$$\vec{B} = \mu_0 \mu(\omega) \vec{H} .$$

By using the boundary conditions of electromagnetic fields at the transition between two media

$$\begin{aligned} \vec{N} \times (\vec{E}_2 - \vec{E}_1) &= 0 , \\ \vec{N} \times (\vec{H}_2 - \vec{H}_1) &= 0 , \\ \vec{N} \cdot (\vec{D}_2 - \vec{D}_1) &= 0 , \\ \vec{N} \cdot (\vec{B}_2 - \vec{B}_1) &= 0 , \end{aligned}$$

---

[3]The established name for such man-made composites was originally coined by R. M. Walser in 1999 [48]. The word "metamaterial" refers on the Greek term $\mu\varepsilon\tau\alpha$ for *beyond*.

and the Maxwell curl equations (2.15)–(2.16), we can derive the Fresnel equations which in turn link n and Z to the transmittance T and the reflectance R. The numeric indices denote the corresponding medium, whereas $\vec{N}$ is a normalized vector perpendicular to the boundary surface of both media. The resulting reflectances for s- and p-polarization[4] are given by

$$R_s = |r_s|^2 = \left|\left(\frac{E_0^r}{E_0^i}\right)_s\right|^2 = \left|\frac{n_1 \cos\theta_i - \frac{\mu_1}{\mu_2} n_2 \cos\theta_t}{n_1 \cos\theta_i + \frac{\mu_1}{\mu_2} n_2 \cos\theta_t}\right|^2 = \left|\frac{\cos\theta_i - \frac{Z_1}{Z_2}\cos\theta_t}{\cos\theta_i + \frac{Z_1}{Z_2}\cos\theta_t}\right|^2, \quad (2.20)$$

$$R_p = |r_p|^2 = \left|\left(\frac{E_0^r}{E_0^i}\right)_p\right|^2 = \left|\frac{\frac{\mu_1}{\mu_2} n_2 \cos\theta_i - n_1 \cos\theta_t}{\frac{\mu_1}{\mu_2} n_2 \cos\theta_i + n_1 \cos\theta_t}\right|^2 = \left|\frac{\frac{Z_1}{Z_2}\cos\theta_i - \cos\theta_t}{\frac{Z_1}{Z_2}\cos\theta_i + \cos\theta_t}\right|^2, \quad (2.21)$$

respectively, where E_0^r is the reflected and E_0^i the incident electric-field amplitude (for further parameter definitions refer to Fig. 2.1). The equations for the transmittance of each polarization read

$$\begin{aligned} T_s &= \left|\frac{n_2 \cos\theta_t}{n_1 \cos\theta_i}\right| |t_s|^2 = \left|\frac{n_2 \cos\theta_t}{n_1 \cos\theta_i}\right| \left|\left(\frac{E_0^t}{E_0^i}\right)_s\right|^2 \\ &= \left|\frac{n_2 \cos\theta_t}{n_1 \cos\theta_i}\right| \left|\frac{2 n_1 \cos\theta_i}{n_1 \cos\theta_i + \frac{\mu_1}{\mu_2} n_2 \cos\theta_t}\right|^2 \end{aligned} \quad (2.22)$$

$$\begin{aligned} T_p &= \left|\frac{n_2 \cos\theta_t}{n_1 \cos\theta_i}\right| |t_p|^2 = \left|\frac{n_2 \cos\theta_t}{n_1 \cos\theta_i}\right| \left|\left(\frac{E_0^t}{E_0^i}\right)_p\right|^2 \\ &= \left|\frac{n_2 \cos\theta_t}{n_1 \cos\theta_i}\right| \left|\frac{2 n_1 \cos\theta_i}{n_2 \frac{\mu_1}{\mu_2} \cos\theta_i + n_1 \cos\theta_t}\right|^2 \end{aligned} \quad (2.23)$$

at which E_0^t is the transmitted electric-field amplitude. The Fresnel equations (2.20)–(2.23) are expressed as functions of the complex refractive index

$$n^2(\omega) = \mu(\omega)\varepsilon(\omega) \quad (2.24)$$

and the complex impedance

$$Z(\omega) = \frac{|\vec{E}|}{|\vec{H}|} = \sqrt{\frac{\mu_0 \mu(\omega)}{\varepsilon_0 \varepsilon(\omega)}} = Z_0 \cdot \sqrt{\frac{\mu(\omega)}{\varepsilon(\omega)}}, \quad (2.25)$$

where $Z_0 = 376.7\,\Omega$ is the vacuum impedance.

It is noteworthy that one can also compute the impedance $Z(\omega)$ and the refractive index $n(\omega)$ of a magneto-dielectric material by inverting the Fresnel equations. For a slab embedded in a homogeneous, surrounding background medium and for normal incidence, this results in [49]

$$n(\omega) = \pm \arccos\left(\frac{1 - r^2 + t^2}{2t}\right)(k_0 d_s)^{-1} + \frac{2\pi m}{k_0 d_s}, \quad (2.26)$$

$$Z(\omega) = \pm Z_{\text{bg}} \sqrt{\frac{(1+r)^2 - t^2}{(1-r)^2 - t^2}}, \quad (2.27)$$

---

[4] p-polarization means that the electric field vector $\vec{E}$ is parallel to the plane of incidence. Accordingly, s-polarization describes the situation where the electric field vector $\vec{E}$ is perpendicular.

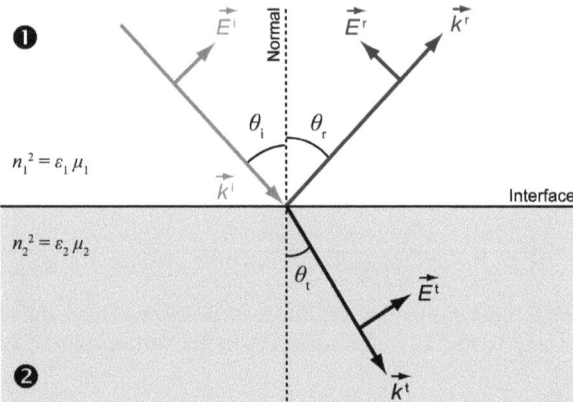

Figure 2.1: *p*-polarized light, represented by its electric field vector, is impinging from medium ❶ onto another medium ❷. At the interface, the incident beam $\vec{E}^i$ splits into a transmitted and a reflected beam ($\vec{E}^t$ and $\vec{E}^r$, respectively). Due to different refractive indices $n_1 \neq n_2$, the reflected and the transmitted beams do *not* include the same angle with the interface's normal ($\theta_t \neq \theta_r$), whereas $\theta_i = \theta_r$.

where d_s denotes the thickness of the material slab, Z_{bg} the impedance of the background medium, and $m \in \mathbb{Z}$. $Z(\omega)$ and $n(\omega)$ enable in turn to retrieve the material parameters from the numerically calculated reflection coefficient r and transmission coefficient t which still include the acquired phase shifts during propagation through the material slab. Note that the knowledge of the intensities $R(\omega) = |r(\omega)|^2$ and $T(\omega) = \left|\frac{n_2 \cos\theta_t}{n_1 \cos\theta_i}\right| |t(\omega)|^2$ alone is not sufficient. Furthermore, the inverted Fresnel equations (2.26)–(2.27) disclose that $n(\omega)$ and $Z(\omega)$ are mathematically not uniquely determined since there exist multiple branches of possible solutions. Thus, an interpretation might lead to ambiguities in the determination of $\varepsilon(\omega)$ and $\mu(\omega)$. We can partly resolve these problems by realizing that at low frequencies, the acquired phase $\varphi(\omega) = \omega n(\omega) d/c_0$ becomes smaller than 2π. This, together with assumptions concerning the material properties (e.g., $\text{Im}(n) > 0, \text{Re}(Z) > 0$ for passive media) sorts out all physically irrelevant solutions.

Next, let us have a look at effects arising from different angles of light incidence. Under normal incidence ($\theta_i = \theta_t = 0°$), (2.20) and (2.21) directly show that reflection occurs if $Z_1 \neq Z_2$. Only for the special case of $\mu(\omega) = 1$ this is equivalent to saying that reflection occurs when $n_1 \neq n_2$. Under oblique incidence $\theta_i \neq 0°$, the Fresnel equations for *p*-polarized light show zero reflectance for the Brewster angle $\theta_{i,p}^B$, whereas the reflectance for *s*-polarization is non-zero over the whole range of angles. If medium 1 is air and medium 2 consists of a non-magnetic material, the Brewster angle is given by

$$\tan(\theta_{i,p}^B) = n(\omega) = \sqrt{\varepsilon(\omega)} \ .$$

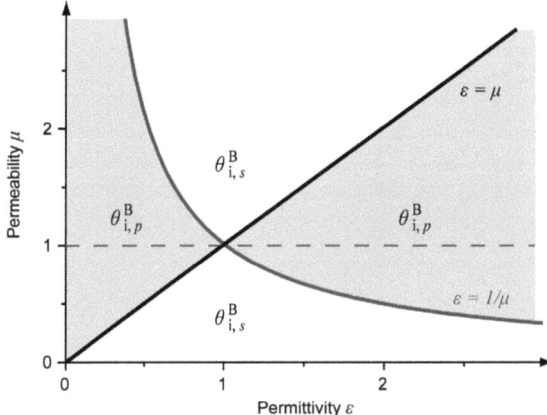

Figure 2.2: Regions of Brewster angles for different polarizations as $\mu(\omega) \neq 1$. Here, μ and ε are supposed to be real values. The regions denoted by $\theta^{\text{B}}_{i,s}$ define values of ε and μ for which the Brewster angle exists only for s-polarization of the incident light field. Regions denoted by $\theta^{\text{B}}_{i,p}$ are for p-polarization. The solid lines indicate where a Brewster angle exists for both polarizations. The dashed line denotes the non-magnetic case ($\mu(\omega) \equiv 1$). Adapted from Ref. [50].

If we, however, permit $\mu(\omega) \neq 1$ for medium 2, the Brewster angles for p- and s-polarization read [50]

$$\tan^2(\theta^{\text{B}}_{i,s}) = \frac{\mu(\omega)\left(\varepsilon(\omega) - \mu(\omega)\right)}{1 - \varepsilon(\omega)\mu(\omega)},$$

$$\tan^2(\theta^{\text{B}}_{i,p}) = \frac{\varepsilon(\omega)\left(\mu(\omega) - \varepsilon(\omega)\right)}{1 - \varepsilon(\omega)\mu(\omega)}.$$

As shown in Fig. 2.2, also a magnetic Brewster angle for s-polarization is expected. Additionally to the electric dipoles, the magnetic dipoles do not emit along the oscillation axis if s-polarized light is impinging under an angle $\theta^{\text{B}}_{i,s}$.

2.4. Design of Magnetic Building Blocks

To practically evoke a magnetic response, one has to generate magnetic dipoles by circulating currents. This can be done, e.g., with ring-shaped metallic unit cells providing an inductance L. The magnetic dipole moment is given by the product of the ring current I and the area A of the coil and is oriented perpendicular to the plane of the coil. At first glance, a structure consisting of ring-shaped elements comes up to our expectations. Unfortunately, the magnetic response of such a composite would be far too small to be useful. Therefore, it is necessary to utilize resonant building blocks like the so-called split-ring resonators (SRRs) [10–12, 24, 25] which embody an additional capacitance C. SRRs consist of wires which are bent to form a coil with one winding ($q.v.$ Fig. 2.3) and a capacitor at the slit. Consequently, each unit cell can be considered as an LC oscillatory circuit having a characteristic eigenfrequency ω_0 at which

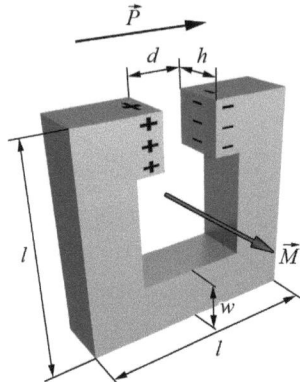

Figure 2.3: Illustration of a split-ring resonator including geometrical parameters. Adapted from Ref. [51].

a resonantly enhanced current flow is expected. That again results in a resonantly enhanced magnetic dipole moment. Of course, in reality, energy is also dissipated by ohmic losses in the metal or can be radiated into free space, leading to the radiation resistance [52, 53]. The effect of both can be merged into one effective resistance R leading to an LCR circuit with eigenfrequency ω_{LCR}. For frequencies above ω_{LCR}, in close analogy to any harmonic oscillator, the response is phase delayed by 180° with respect to the driving force. Hence, the induced magnetic field of the SRR is opposite to the driving magnetic field, resulting in a diamagnetic behavior.

To derive ω_{LCR} for a SRR, we use the quasi-static approximation for the inductance L of a long coil while setting the number of windings to one. This results in

$$L = \frac{\Phi_{\mathrm{B}}(t)}{I(t)} = \mu_0 \frac{l^2}{h} \, , \qquad (2.28)$$

where Φ_{B} is the magnetic flux and I the current inside the coil. The geometry parameters w, h, l, and d are defined in Fig. 2.3. The quasi-static equation of a plate capacitor reads

$$C = \frac{Q(t)}{U(t)} = \varepsilon_0 \varepsilon_{\mathrm{gap}} \frac{wh}{d} \, , \qquad (2.29)$$

with the time-dependent charge $Q(t)$ and voltage $U(t)$. $\varepsilon_{\mathrm{gap}}$ is the permittivity of the material placed inside the gap between both capacitor plates. To obtain all contributions to the voltage induced by the incident light we recall Kirchhoff's second law, i.e.,

$$U_{\mathrm{ind}}(t) = L \frac{\partial I}{\partial t} + \frac{1}{C} \int I \, \mathrm{d}t + RI \, . \qquad (2.30)$$

The eigenfrequency is deduced by setting the right-hand side of (2.30) to zero and differentiating with respect to the time t so that

$$L \frac{\partial^2 I}{\partial t^2} + \frac{1}{C} I + R \frac{\partial I}{\partial t} = 0 \, .$$

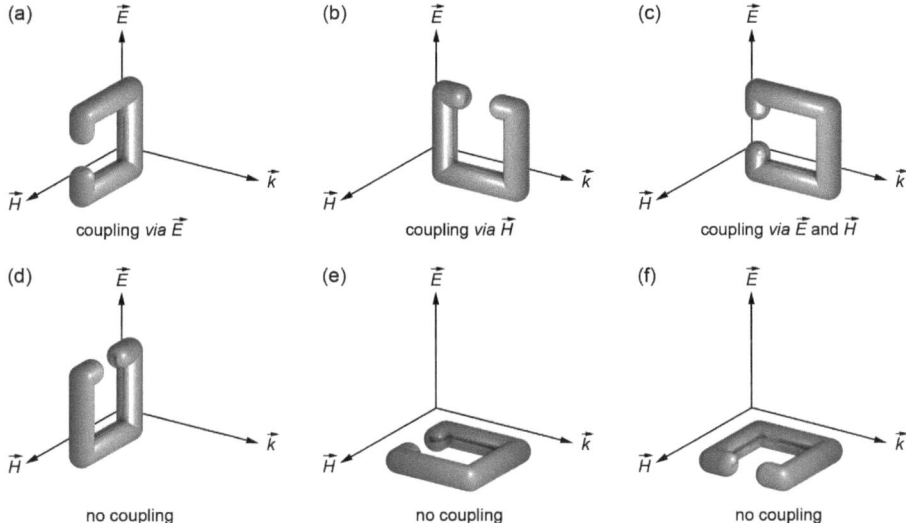

Figure 2.4: Illustration of a SRR excited in different configurations (only linear-polarized light). For **(a)**, coupling to the LCR resonance is only possible *via* the electric field and for **(b)** only *via* the magnetic field. For the case **(c)**, both electromagnetic field components can couple to the magnetic resonance. For the other configurations **(d)**, **(e)**, and **(f)**, however, it is not possible to couple to the magnetic resonance at all.

By using the common ansatz $I(t) = I_0\,e^{-i\omega t}$, we obtain the characteristic differential equation of a damped harmonic oscillator

$$-\omega^2 I(t) + \frac{1}{LC} I(t) - i\omega \frac{R}{L} I(t) = 0 \;. \tag{2.31}$$

By replacing the factors $1/(\sqrt{LC}) := \omega_0$ (i.e., the eigenfrequency of an undamped LC circuit) and $R/(2L) := \gamma$ (i.e., the damping factor), we end up with a quadratic equation

$$-\omega^2 - 2i\gamma\omega + \omega_0^2 = 0$$

which can be solved by

$$\begin{aligned}\omega_{LCR} &= -\frac{1}{2}\left(2i\gamma \pm \sqrt{-4\gamma^2 + 4\omega_0^2}\right) \\ &= -i\gamma \pm \sqrt{\omega_0^2 - \gamma^2}\;. \end{aligned} \tag{2.32}$$

Clearly, for low damping $\gamma \ll \omega_0$, the eigenfrequency of the LCR circuit approaches the LC limit, i.e., $\omega_{LCR} \approx \omega_0$.

To derive $\mu(\omega)$ of a periodic array of SRRs, we specify a certain excitation geometry shown in Fig. 2.4(b) which enables a purely magnetic coupling in the quasi-static limit, i.e., retardation effects can be neglected. The spatial directions of the considered vectors are obvious so that

15

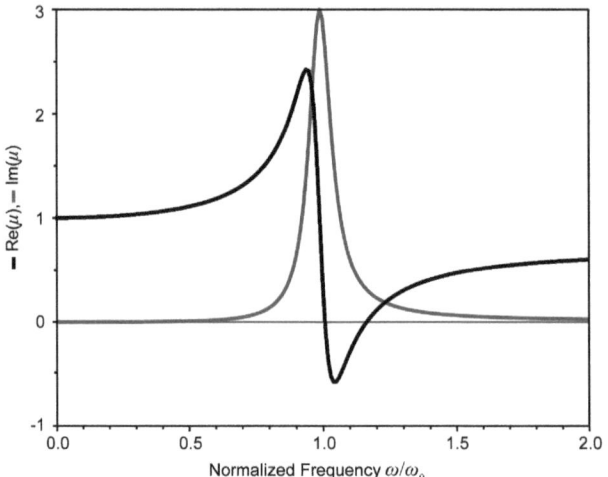

Figure 2.5: Dispersion of $\mu(\omega)$ of a SRR array plotted *versus* the normalized frequency. The values are calculated by using (2.34) with the parameters $f = 0.3$ and $\gamma/\omega_0 = 0.05$. One obtains a Lorentzian-like progression which also leads to negative values of Re(μ) (black) for $1 < \omega/\omega_0 < 1.2$.

the calculations can be reduced to absolute values. The derivation starts again with the ansatz (2.30) by assuming a homogeneous magnetic field in the SRR coil. In this case, the magnetic flux is given by $\Phi_B = \mu_0 l^2 H$, where H is the exciting time-harmonic magnetic field. Proceeding similarly to (2.31) leads to

$$\frac{\partial U_{\text{ind}}}{\partial t} = -\frac{\partial^2 \Phi_B}{\partial t^2} \Rightarrow \frac{\partial^2 I}{\partial t^2} + \frac{1}{LC} I + \frac{R}{L} \frac{\partial I}{\partial t} = \omega^2 \frac{\mu_0 l^2}{L} H \ .$$

With simple mathematical operations we obtain

$$H = \left(\frac{\omega_0^2 - \omega^2 - 2i\gamma\omega}{\mu_0 l^2 \omega^2} \right) IL \ . \tag{2.33}$$

In (2.10), we found that the magnetization M and the external magnetic field H are related via the magnetic susceptibility $\chi_m(\omega) = \mu(\omega) - 1$. The contribution of each SRR to the total magnetization is given by

$$M = \frac{m}{V} = \frac{Il^2}{V} \ ,$$

where V denotes the unit volume. For a periodic arrangement of SRRs, the latter can be expressed by the respective periodicities a_i of the lattice in each spatial direction, i.e., $V = \prod_{i=1}^{3} (a_i)$.

Finally, $\mu(\omega)$ of the considered excitation geometry is given by

$$\begin{aligned}
\mu(\omega) &= 1 + \frac{M}{H} \\
&\stackrel{(2.33)}{=} 1 + \frac{Il^2}{V} \frac{\mu_0 l^2 \omega^2}{(\omega_0^2 - \omega^2 - 2i\gamma\omega)IL} \\
&\stackrel{(2.28)}{=} 1 + \frac{l^2 h}{V} \left(\frac{\omega^2}{\omega_0^2 - \omega^2 - 2i\gamma\omega} \right) \\
&= 1 + \left(\frac{f\omega^2}{\omega_0^2 - \omega^2 - 2i\gamma\omega} \right) ,
\end{aligned} \qquad (2.34)$$

where we lumped the prefactors to a filling fraction f with

$$0 \leq f = \frac{l^2 h}{V} = l^2 h \prod_{i=1}^{3} \left(\frac{1}{a_i} \right) \leq 1 . \qquad (2.35)$$

Note that $f=1$ corresponds to the case where all SRRs are touching each other, i.e., the upper bound of an obtainable SRR density. Fig. 2.5 shows the dispersion of the permeability plotted for the parameters $f = 0.3$ and $\gamma/\omega_0 = 0.05$. Roughly speaking, (2.34) represents a Lorentz oscillator (for details see section A.3). A subtle difference is the ω^2 numerator, which leads to the asymptotics $\mu(0)=1$ and $\mu(\infty)=1-f$. In the static limit, no current can be induced which is also reproduced by our quasi-static model. The behavior for infinite frequencies is rather a symptom of the model and would have to be replaced by $\mu(\infty) = 1$ when accounting for a Drude-type metal (to be discussed in section 2.6.2).

Further improvements of the SRR model have been developed [54–56] to bring its quantitative behavior closer to calculations and experiments. By way of example, the inductive contributions have been separated to a kinetic and a magnetic part, $L_{\text{kin}} \sim d/(wh)$ and L_{mag}, respectively. Importantly, L_{kin} reintroduces a thickness dependence of the eigenfrequency, i.e., the thicker the SRR the higher ω_{LCR}. Moreover, the SRR's capacitance can be separated to a surface and a gap contribution.

Note that the magnetic field $B \sim hE/\lambda \ll E$ of the capacitor and the electric field $E \sim lE/\lambda \ll B$ of the coil are neglected [57] in the quasi-static approximation. These assumptions are only justified if the wavelength λ is much larger than the thickness h of the capacitor and the length l of the coil. However, we will see for a special example in section 4.1 that the qualitative behavior is fairly well reproduced. And in any case, the model gives an indication on how to tune the geometrical parameters in order to modify the spectral position of ω_{LCR}.

Another magnetic unit cell results by further increasing the slit width d of the SRR. The increased gap lowers the capacitance and hence increases the LCR eigenfrequency. Introducing a second gap at the bottom arm of the U-shaped wire leads to a second serial capacitance further reducing the net capacitance. The resulting structure (see Fig. 2.6(a)) is named "cut-wire pair" [58, 60]. Additionally tilting by 90° results in the unit cell shown in Fig. 2.6(b). A consequence of this transition is the replacement of ohmic currents in the horizontal arms of the SRR by displacement currents.

The cut-wire pair can also be interpreted as a pair of coupled antennas (oscillators) which exhibit two eigenmodes (shown in Fig. 2.6(b)). On the one hand, an anti-symmetric mode is

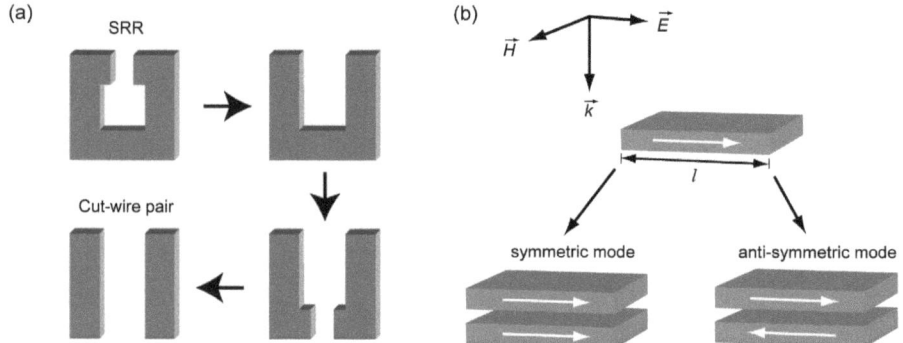

Figure 2.6: Cut-wire pairs can be considered as a geometrically modified split-ring resonator (SRR) for which the electrical connection between both metal plates is replaced by a displacement current. (a) Schematic of the transition from SRRs (top left) to cut-wire pairs (bottom left). Adapted from Ref. [58]. (b) Another intuitive picture for the cut-wire pair can be derived from an antenna which is excited by a plane wave (resonance wavelength of the antenna: $\lambda = l/2$). Adding another antenna leads to a coupling effect which results in two eigenmodes. The anti-symmetric mode (right) exhibits a magnetic dipole, while the symmetric mode (left) is related to an electric resonance. Adapted from Ref. [59].

present with current oscillations in both antennas opposite in phase. On the other hand, one can also observe a symmetric mode with in-phase oscillations. The magnetic dipole moment of the symmetric mode is, however, negligible compared to the anti-symmetric mode. Obviously, the origin of the induced magnetic dipole moment is based on retardation effects of the incident light wave. Strictly speaking, this is contradictory to treating cut-wire pairs as unit cells of an effective medium. Referring to the last paragraph of section 2.1, this is a typical example of a "black-boxed" metamaterial as we refrain from analyzing the microscopic reasons for the magnetic response but rather map them to mesoscopic material parameters.

Other derivatives of the SRR are shown in Fig. 2.7. One clearly perceives that all structures consist of inductive and capacitive elements turning them into LCR circuits. Thus, the formalism to describe these magnetic elements is roughly similar to SRRs.

In summary, by using SRRs (and derivatives) as magnetic constituents of a composite medium, we are able to mimic a permeability unequal to one at certain frequency intervals as there exists an eigenfrequency ω_{LCR} at which a resonant current can be driven. In the following sections 2.5–2.8, a variety of ideas in view of applications will be presented which are subdivided into (i) purely magnetic structures ($\text{Re}(\mu) \neq 1$), (ii) negative-index materials ($\text{Re}(\mu), \text{Re}(\varepsilon) < 0 \Rightarrow \text{Re}(n) < 0$), and (iii) bi-anisotropic structures.

2.5. Purely Magnetic Metamaterials

Zero reflection at an interface of two media is observed if both impedances are perfectly matched. For most real-world applications (e.g., cameras, wafer steppers, projectors), one of those media

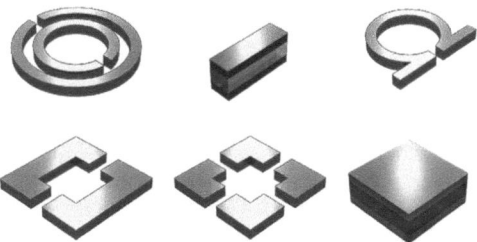

Figure 2.7: Illustration of a variety of different magnetic building blocks deduced from the SRR shown in Fig. 2.3. The magnetic response of all presented unit cells is based on a resonant LCR oscillatory circuit. Adapted from Ref. [7].

is air ($\varepsilon(\omega) = \mu(\omega) \approx 1$). By looking at the definition of the impedance Z given by (2.25), perfect matching of two isotropic materials is, thus, only possible for $\varepsilon(\omega) \equiv \mu(\omega)$. Assuming that we know how to fabricate a composite with arbitrarily high values of ε and μ, it is conceivable to realize non-reflecting lenses providing a high refractive index. Notably, unlike $Z(\omega)$, the refractive index $n(\omega)$ is determined by the product of ε and μ (q.v. (2.24)).

Such highly refracting metamaterials would be of immense technological interest, since they could be used for tiny and efficient objectives. Unfortunately, to date there exists no proposal on how to achieve extraordinarily high but still identical values for ε and μ. Besides that, metal has to be used to obtain a magnetic response. That again accounts for substantial absorption losses due to non-zero imaginary parts of the material parameters.

2.6. Negative-Index Metamaterials

In 1968, Victor G. Veselago predicted the optical behavior of "hypothetic" materials with an intrinsic negative real part of the refractive index [9]. A sufficient condition to attain this property is met if the real parts of both ε and μ are simultaneously negative (to be derived in the following section). As mentioned, $\text{Re}(\mu) < 0$ had not been observed for natural materials. Hence, the concept fell into oblivion until technological progresses enabled the realization of magnetic unit cells on the micro- and nanoscale leading to man-made composite materials [12, 14, 25]. Since then, negative-index metamaterials evoked a lot of interest in the field of photonics as they form a new subclass of optical materials.

A typical characteristic of a "left-handed" medium[5] is the refraction of light to the "wrong" side as shown in Fig. 2.8. However, negative refraction can also be obtained from photonic crystals due to interference of partial waves from different lattice points [61–65], by non-magnetic meta-

---

[5]The term "left-handed" was used due to the fact that the Poynting vector $\vec{S}$ of light inside a negative-index material is anti-parallel to the wave vector $\vec{k}$ (see Fig. 2.8). Consequently, $\vec{S}$, $\vec{E}$, and $\vec{B}$ form a left-handed system. In the context of this Thesis, we will refrain from using this terminology since "handedness" is rather related to optical activity (chirality) and might be confusing. Therefore, it is appropriate to use the term "negative-index" material.

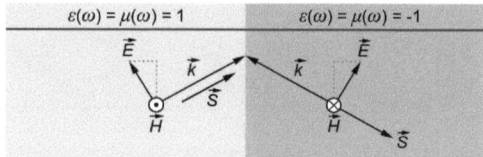

Figure 2.8: Illustration of negative refraction at the interface between air half-space ($\varepsilon(\omega)=\mu(\omega)=1$) on the left-hand side and a negative-index half-space with $\varepsilon(\omega)=\mu(\omega)=-1$ on the right-hand side for p-polarization of the incident light. On the right-hand side, $n(\omega)=-1$. Also note that the wave vector $\vec{k}$ (solid) and the energy flux (Poynting vector) $\vec{S}=\frac{1}{2}\mathrm{Re}(\vec{E}\times\vec{H}(c.c.))$ of light inside the negative-index material are anti-parallel to each other.

materials with hyperbolic dispersion [66], by anisotropic crystals at certain angles of incidence, and even by homogeneous isotropic metal layers [67]. Thus, it is important to carefully distinguish between a "negative refractive index" and "negative refraction" since the underlying physics can be fundamentally different.

Negative group velocities [68] and negative phase velocities [68–70] have also been investigated for negative-index materials but are—similar to the negative refraction—not unique indicators since they also appear for particular dispersive media and surface plasmon polaritons [71]. Indeed, some remarkable effects can be exclusively attributed to negative-index metamaterials and have never been found for natural substances, i.e., the inverse Doppler effect [17], the negative Goos-Hänchen shift [69, 72, 73], and reversed Čerenkov radiation [18, 19].

2.6.1. Passive Medium Conditions

Before we discuss appropriate structure designs, it is crucial to figure out what kind of sufficient and necessary conditions must be fulfilled to obtain a negative real part of the refractive index for passive media. "Passive" means that the considered medium is incapable of amplifying an incident electromagnetic wave. For clarity, we will restrict our discussions to the isotropic case. Thus, the vector properties of the fields can be replaced by their absolute values. Notably, the conditions to be derived keep being valid for the general bi-anisotropic case (see section 2.8).

(i) The imaginary part of the refractive index $\mathrm{Im}(n) := n_{\mathrm{im}}$ is positive definite for a passive medium. We constitute this statement by using the dispersion relation of an isotropic medium $k = n\omega/c_0 = nk_0$, where k is the complex wave number inside the medium and k_0 the wave number in vacuum. Next, this dispersion relation is combined with the electric field of an harmonic plane wave $E(r) = E_0\, e^{ikr}$ and the refractive index is separated to its real and imaginary parts $n(\omega) = n_{\mathrm{re}} + in_{\mathrm{im}}$, respectively. This results in

$$E = E_0\, e^{ik_0 n_{\mathrm{re}} r}\, e^{-k_0 n_{\mathrm{im}} r}\,. \qquad (2.36)$$

Equation (2.36) clearly reveals that $n_{\mathrm{im}} < 0$ causes an exponential increase of the amplitude with growing distance which is an unphysical solution for passive media. We also identify n_{im} to be a measure for the absorption of the medium.

Note that it is common to introduce the negative ratio of real and imaginary parts of n as a measure for the performance of negative-index materials. Hence, the "Figure of Merit" (FOM) is defined as
$$\text{FOM} := -\frac{n_{\text{re}}}{n_{\text{im}}} . \tag{2.37}$$
We obtain high *positive* values for the FOM if the metamaterial features high *negative* values for n_{re} while keeping the absorption relatively low.

(ii) The real part of the impedance Re(Z) is positive definite. This can be derived by using an infinitesimally-thin conductive sheet which is infinitely elongated in y- and z-direction and located at $x = x_0$ (see Fig. A.1). A time-varying current density $j = j_0 e^{-i\omega t}$ is induced in this sheet by an external monochromatic harmonic wave. Thus, the 1D wave equation is given by [74]
$$\frac{\partial^2 E(x,t)}{\partial x^2} + \frac{\omega^2 n^2}{c_0^2} \frac{\partial^2 E(x,t)}{\partial t^2} = -i\omega\mu_0\mu j_0\, e^{-i\omega t}\delta(x-x_0) , \tag{2.38}$$
where we assume, for simplification, that the material parameters are constant in frequency. At the position $x = x_0$, (2.38) can be solved by
$$E(x,t) = -\frac{c_0\mu_0\mu}{2n}\, j_0\, e^{i(nk|x-x_0|-\omega t)} . \tag{2.39}$$
A detailed mathematical proof can be found in section A.4. To increase entropy, the external wave must in average do positive work on the complex fields. Hence, the temporally averaged power has to be greater than zero, i.e.,
$$\begin{aligned}\langle P\rangle_t &= \frac{1}{T}\int_0^T P(t)\,\mathrm{d}t = \langle U(t)I(t)\rangle_t \\ &= \left\langle \int_r E(t)\,\mathrm{d}r' \int_A j(t)\,\mathrm{d}A' \right\rangle_t \stackrel{Ar\,\equiv\,V}{=} -\frac{1}{2}\int_V j^* E(x,\omega)\,\mathrm{d}x \\ &\stackrel{(2.39)}{=} \frac{c_0\mu_0\mu}{4n} j_0^2 \stackrel{(2.25)}{=} \frac{1}{4}Z j_0^2 \stackrel{!}{>} 0 .\end{aligned} \tag{2.40}$$
If we also take dispersion of the material parameters into account, the electric field has a similar appearance as in (2.39) [74], and the resulting averaged power, in analogy to (2.40), is given by
$$P(\omega) \sim Z(\omega)\,|j(\omega)|^2 \stackrel{!}{>} 0 . \tag{2.41}$$
Thus, as we restrict our discussion on passive media, Re($Z(\omega)$) must be positive definite to fulfill the inequality.

(iii) The energy density inside a dispersionless medium is given by
$$U = \varepsilon E^2 + \mu H^2 . \tag{2.42}$$
If both permittivity and permeability are simultaneously negative, U will be negative which contradicts thermodynamic laws. Apparently, we have to replace (2.42) by a relation which accounts for dispersion, i.e., [9, 35, 75]
$$U = \text{Re}\left(\frac{\partial(\varepsilon\omega)}{\partial\omega}\right)|\vec{E}|^2 + \text{Re}\left(\frac{\partial(\mu\omega)}{\partial\omega}\right)|\vec{H}|^2 ,$$

whereas the law of increase of entropy requires the conditions

$$\text{Re}\left(\frac{\partial(\varepsilon\omega)}{\partial\omega}\right), \text{Re}\left(\frac{\partial(\mu\omega)}{\partial\omega}\right) > 0$$

to be fulfilled. These inequalities do not prohibit simultaneous negative values for ε and μ, but rather enforce a frequency dependence.

(iv) We expect $\varepsilon(\omega) = \varepsilon_{\text{re}} + i\varepsilon_{\text{im}}$ and $\mu(\omega) = \mu_{\text{re}} + i\mu_{\text{im}}$ to be continuous smooth functions. Neither discontinuities in these functions nor in their spectral derivatives are expected.

By knowing these conditions, we are now prepared to discuss n_{re} starting from (2.24):

$$n^2 = \varepsilon(\omega)\mu(\omega)$$
$$\Leftrightarrow n_{\text{re}}^2 + 2in_{\text{re}}n_{\text{im}} - n_{\text{im}}^2 = \varepsilon_{\text{re}}\mu_{\text{re}} + i\varepsilon_{\text{re}}\mu_{\text{im}} + i\varepsilon_{\text{im}}\mu_{\text{re}} - \varepsilon_{\text{im}}\mu_{\text{im}} \ . \tag{2.43}$$

The real and imaginary parts in (2.43) can be separated to

$$\text{real} : \quad n_{\text{re}} = \pm\sqrt{\varepsilon_{\text{re}}\mu_{\text{re}} - \varepsilon_{\text{im}}\mu_{\text{im}} + n_{\text{im}}^2} \ , \tag{2.44}$$
$$\text{imaginary} : \quad 2in_{\text{re}}n_{\text{im}} = i\varepsilon_{\text{re}}\mu_{\text{im}} + i\varepsilon_{\text{im}}\mu_{\text{re}} \ . \tag{2.45}$$

Since (2.44) results again in a quadratic equation, we rather continue with (2.45). If we recall the conditions (i)–(iii), the only possibility is given for

$$n_{\text{re}} = \frac{\varepsilon_{\text{re}}\mu_{\text{im}} + \varepsilon_{\text{im}}\mu_{\text{re}}}{2n_{\text{im}}} \stackrel{!}{<} 0 \Rightarrow \varepsilon_{\text{re}}\mu_{\text{im}} + \varepsilon_{\text{im}}\mu_{\text{re}} < 0 \ . \tag{2.46}$$

The inequality in (2.46) is fulfilled if, e.g., the real or imaginary parts of both ε and μ are simultaneously negative. Due to the intermixture of real and imaginary parts, these are sufficient but not necessary conditions.

2.6.2. Drude Model and Diluted Metals

Previously, we have seen that the dispersion of the permeability $\mu(\omega)$ of a magnetic unit cell can be described by a slightly modified Lorentz oscillator model. Hence, at the resonance and for low damping, it is possible to achieve $\text{Re}(\mu) < 0$. Next, we have to find ways to attain a negative permittivity so that (2.46) is satisfied. Here, nature helps by providing the dispersion of metals. In standard textbooks on solid-state physics (e.g., [2, 3]) the classic description of metals is given by Drude's free electron model. As this model assumes that electrons do not feel any restoring force, we use the result of the Lorentz oscillator model (A.5) without the restoring term $\sim \omega_j^2$. Thus, the permittivity of the Drude model reads

$$\varepsilon(\omega) = 1 - \frac{\omega_{\text{pl}}^2}{\omega^2 + i\gamma\omega} \ , \tag{2.47}$$

where $\omega_{\text{pl}} = \sqrt{(n_0 e^2)/(m_{\text{eff}}\varepsilon_0)}$ is the plasma frequency, n_0 the electron density, and γ the damping factor. The effective mass m_{eff} is given by the curvature of the conduction-band dispersion and corrects the model for the periodic potential of a crystal lattice [2, 3].

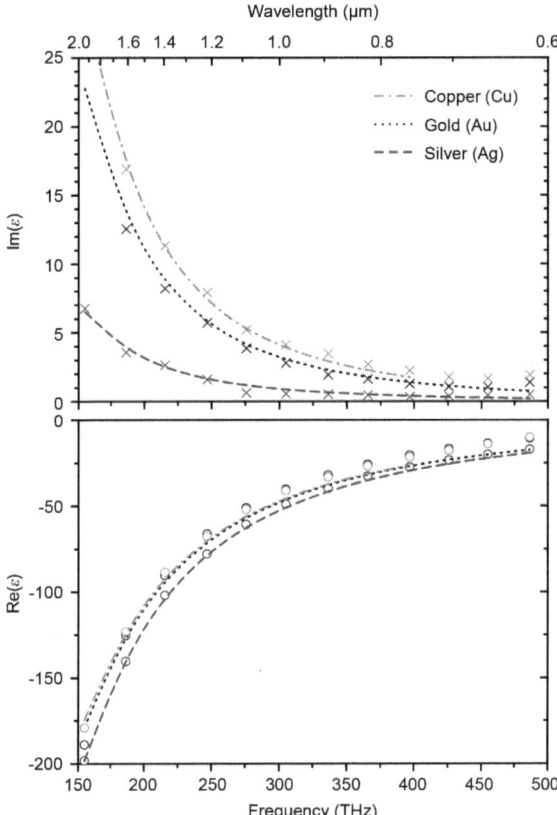

Figure 2.9: Dispersion of the permittivity $\varepsilon(\nu)$ of silver (dashed line), gold (dotted line), and copper (dash-dotted line) at near-infrared to visible frequencies between 150 THz (2.0 μm wavelength) to 500 THz (0.6 μm wavelength). The circles and crosses denote the experimentally determined [76] Re(ε) and Im(ε), respectively. Note that thin films were used for the measurements with thicknesses of 30.4 nm and 37.5 nm for silver, 34.3 nm and 45.6 nm for gold, as well as 29.7 nm and 30.5 nm for copper. The films have been prepared by using vacuum evaporation which results in very smooth metal layers with a thickness error of ±0.2 nm. The lines correspond to Drude fits using (2.47). The curve fit for copper is only performed up to 400 THz since appearing interband transitions would influence the Drude parameters.

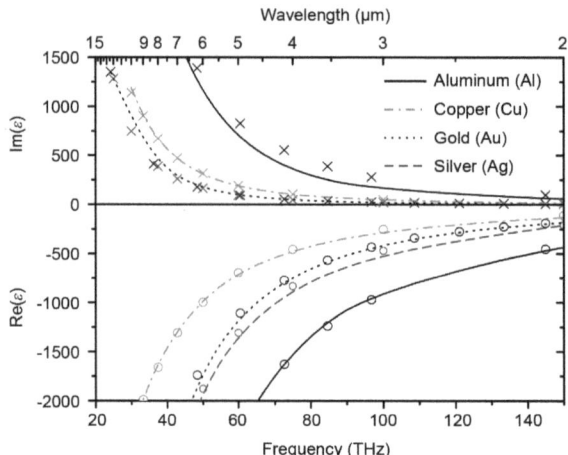

Figure 2.10: Relative permittivity $\varepsilon(\nu)$ of silver (dashed line), gold (dotted line), copper (dash-dotted line), and aluminum (solid line) at mid-infrared frequencies between $\nu = \omega/(2\pi) = 20\,\text{THz}$ (15 µm wavelength) to 150 THz (2 µm wavelength). The circles and crosses denote the experimental data points [77] of $\text{Re}(\varepsilon)$ and $\text{Im}(\varepsilon)$, respectively. The lines correspond to Drude fits using (2.47). Note that the fitting curve of $\text{Im}(\varepsilon)$ of silver is hidden behind that of gold. Clearly, $\text{Im}(\varepsilon)$ of aluminum shows considerable deviations from the Drude model since interband transitions are not considered.

Although the Drude model does not account for interband transitions, it still delivers a good approximation for the permittivity of some noble metals like silver, iridium, gold, and copper up to frequencies of around 400 THz (which corresponds to wavelengths of around 750 nm). By way of example, the interband absorption from the $3d$ to the $4s$ band of copper starts at a frequency of about 535 THz (i.e., a wavelength of 560 nm) which causes perceptible deviations of the experiment [76] from the Drude equation (2.47). In contrast, the interband absorption edge of silver is found at approximately 970 THz (relating to a wavelength of 310 nm) so that the Drude model converges much better to measured data points [76] (see Fig. 2.9). In Fig. 2.10, the dispersion of the permittivity $\varepsilon(\omega/2\pi) = \varepsilon(\nu)$ of aluminum, copper, silver, and gold is shown for the mid-IR. Drude parameters ω_{pl} and γ for the IR/VIS resulting from Drude fits shown as solid lines in Figs. 2.9 and 2.10 are listed in Tab. 2.2. Even though the Drude parameters of Refs. [76, 77] deviate quantitatively due to different sample preparation and spectral region, a qualitative trend of γ can be identified. Apparently, silver and gold have the lowest damping factors. Hence, these noble metals do not suffer that much from ohmic losses as other transition metals do. That makes them preferable for applications like multi-layered metamaterials requiring a high degree of transparency.

Even if the damping is reduced, one still has to face the problem of an impedance mismatch. In the spectral region where photonic metamaterials are expected to show a magnetic response, i.e., at around 100 THz to 300 THz, silver and gold have rather large negative values for $\text{Re}(\varepsilon)$ (q.v. Fig. 2.10). Tying in with the arguments for impedance matching in section 2.5, we require $\varepsilon \approx \mu$ to reduce reflections at the air-metamaterial interface. This means that either

Table 2.2.: Plasma frequency ω_{pl} and damping factor γ obtained by Drude fits (2.47) of experimental data points. The horizontal bar separates the results for the near- [76] and mid-infrared [77] spectral regions. Clearly, the Drude parameters deviate a bit. Anyway, one finds the unique trend of lowest damping γ for silver followed by gold.

Element	Ref.	ω_{pl} (10^{15} Hz)	γ (10^{12} Hz)
Copper (Cu)	[76]	13.06	159.3
Gold (Au)	[76]	13.15	123.2
Silver (Ag)	[76]	13.75	31.8
Aluminum (Al)	[77]	18.70	108.9
Copper (Cu)	[77]	10.22	96.0
Gold (Au)	[77]	12.76	30.2
Silver (Ag)	[77]	13.68	27.4

the magnetic resonance of the metamaterial must deliver large negative values of $\mathrm{Re}(\mu)$ or the electric dipole density inside the metal has to be reduced. The latter can be accomplished by replacing closed metal layers by elongated wires [78]. If they are arranged periodically along the incident electric field vector while the periodicity being much smaller than the wavelength, the structure will behave like a "diluted" metal. Consequently, it is possible to red-shift the plasma frequency ω_{pl} by many orders of magnitude which in turn shifts the absolute values of $\mathrm{Re}(\varepsilon)$. Indeed, this effect is also used for microwave ovens. Small holes in a metal film enable to look through the window, whereas the microwave radiation "feels" a diluted metal which cannot be penetrated.

2.7. Design of Negative-Index Metamaterials

Sections 2.4 and 2.6.2 dealt with the realization of negative permeability and negative permittivity, respectively. Naively thinking, one could simply combine magnetic and diluted metal elements to form a negative-index material. Of course, this seems like an oversimplified approach since coupling effects tend to deteriorate each element's property. Amazingly, it turned out that this method actually works very well since for some geometrical configurations the expected interaction is negligible as both elements act indeed independently. In Fig. 2.11, some exemplary compositions leading to negative-index metamaterials are shown. For example, one can merge SRRs and a diluted metal like shown in Fig. 2.11(a) [12]. Likewise, the double-fishnet structure in Fig. 2.11(b) consists of an array of cut-wire pairs and a metal grid [29–31, 79, 80].

Regrettably, there do not exist any deterministic methods to design negative-index metamaterials. Hence, most of the designs discussed in literature have been found by "educated guessing" or minor modifications of existing structures. Arising changes in the optical properties have mostly been evaluated *via* numerical simulations.

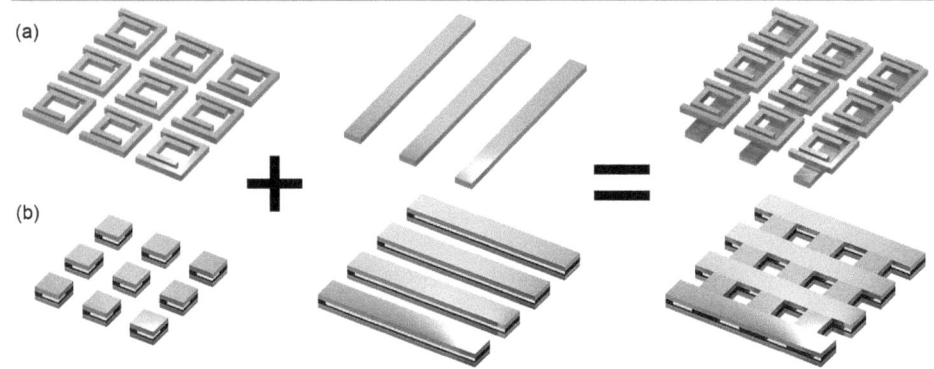

Figure 2.11: "Addition" of magnetic elements on the left-hand side to diluted metal elements on the right-hand side results in negative-index metamaterials. Periodic arrangement of **(a)** SRRs and **(b)** cut-wire pairs combined with wire grids.

2.8. Bi-Anisotropic Metamaterials

For reasons of clarity, the previous discussions were mainly focused on the physical behavior of isotropic metamaterials. Unfortunately, for real-world composites, this formalism is not always flexible enough since some fabricated structures are at least anisotropic (ε and μ are tensors) or even have no centrosymmetry along the propagation direction. For the latter case of general bi-anisotropy, the induced field components include an angle φ with the vectors which excited them. Since the treatment of arbitrarily chosen φ is difficult to handle, we will restrict our considerations to the special cases of $\varphi = 90°$ (i.e., "pure" bi-anisotropy) and $\varphi = 0°$ (i.e., chirality) [81].

2.8.1. "Pure" Bi-Anisotropy

Without loss of generality, we choose a coordinate system in which linearly polarized light propagates along the principle x-axis ($\vec{k} = k_x$). For $\varphi = 90°$, the constitutive relations in (2.11) and (2.12) reduce to [51, 82, 83]

$$D_y = \varepsilon_0 \varepsilon_{yy} E_y - \frac{i}{c_0} \xi_{yz} H_z \,, \tag{2.48}$$

$$B_z = \mu_0 \mu_{zz} H_z + \frac{1}{c_0} \zeta_{zy} E_y = \mu_0 \mu_{zz} H_z + \frac{i}{c_0} \xi_{yz} E_y \,, \tag{2.49}$$

where we defined the material parameters as

$$\underline{\varepsilon} = \begin{pmatrix} \varepsilon_{xx} & 0 & 0 \\ 0 & \varepsilon_{yy} & 0 \\ 0 & 0 & \varepsilon_{zz} \end{pmatrix}, \ \underline{\mu} = \begin{pmatrix} \mu_{xx} & 0 & 0 \\ 0 & \mu_{yy} & 0 \\ 0 & 0 & \mu_{zz} \end{pmatrix}, \text{ and } \underline{\xi} = \begin{pmatrix} 0 & 0 & 0 \\ 0 & 0 & -i\xi_{yz} \\ 0 & 0 & 0 \end{pmatrix}.$$

Equations (2.48)–(2.49) reveal that the electric component of the incident light E_y induces a parallel electric displacement D_y as well as a perpendicular magnetic field B_z. Likewise, the magnetic component H_z induces a parallel magnetic induction and a perpendicular electric displacement. Therefore, the incident linear polarization of the wave is maintained.

To learn more about the optical behavior of pure bi-anisotropic media, we insert (2.48)–(2.49) into Maxwell's curl equations (2.15)–(2.16) to obtain [84]

$$ik_x E_y = i\omega B_z = i\omega \left(\mu_0 \mu_{zz} H_z + \frac{i}{c_0} \xi_{yz} E_y \right) \tag{2.50}$$

$$ik_x H_z = i\omega D_y = i\omega \left(\varepsilon_0 \varepsilon_{yy} E_y - \frac{i}{c_0} \xi_{yz} H_z \right) . \tag{2.51}$$

By additionally using the definition (2.25), the impedance for this special bi-anisotropic case results in

$$Z_{\mathrm{ba}} = Z_0 \left(\frac{\mu_{zz}}{k_x/k_{x,0} - i\xi_{yz}} \right) . \tag{2.52}$$

Apparently, Z_{ba} depends explicitly on the wave vector k_x and, thus, on the propagation direction of the incident light wave. Furthermore, from (2.51) we derive

$$\begin{aligned}
k_x &= \omega \left(\varepsilon_0 \varepsilon_{yy} Z_{\mathrm{ba}} - \frac{i}{c_0} \xi_{yz} \right) \\
&= k_{x,0} \left(\varepsilon_{yy} \frac{Z_{\mathrm{ba}}}{Z_0} - i\xi_{yz} \right) \\
\Rightarrow \frac{k_x}{k_{x,0}} &\stackrel{(2.52)}{=} \frac{\varepsilon_{yy} \mu_{zz}}{k_x/k_{x,0} - i\xi_{yz}} - i\xi_{yz} \\
\Rightarrow k_x^2 &= k_{x,0}^2 \left(\varepsilon_{yy} \mu_{zz} - \xi_{yz}^2 \right) .
\end{aligned} \tag{2.53}$$

Equation (2.53) delivers the expression for the refractive index, i.e.,

$$n_{\mathrm{ba}}^2 = \varepsilon_{yy} \mu_{zz} - \xi_{yz}^2 . \tag{2.54}$$

In contrast to Z_{ba}, no direction dependence results—which would be different in the case of a non-reciprocal medium. As already discussed in section 2.6.1, only the solution with positive imaginary part of n_{ba} is physically relevant. From (2.54) follows that the cross-coupling parameter ξ_{yz} has a high influence on the refractive index and, thus, also on the propagation of waves inside bi-anisotropic media.

To obtain a coarse qualitative overview, we assume a negligible imaginary part of n_{ba} so that $\mathrm{Re}(n_{\mathrm{ba}}^2) > 0$ directly leads to a complex phase of the field equation (2.36) and, therewith, to propagating modes. For the isotropic case, i.e., $\mathrm{Re}(\xi_{yz})=0$, as a rule of thumb light can freely propagate in dielectrics (gray area in Fig. 2.12(a) and (b)) and negative-index materials (black area). $\mathrm{Re}(\xi_{yz}) \neq 0$ leads to a more intricate situation since even for simultaneously positive (or negative) permittivity and permeability, $\mathrm{Re}(n_{\mathrm{ba}}^2)$ is not necessarily positive. Fig. 2.12(b) depicts parameter sets which allow for propagating modes as darkened cone-like volume fractions. Notably, the feasible parameter range for dielectrics and negative-index materials is tremendously decreased. For example, if we assume a medium to be bi-anisotropic—like illustrated in Fig. 2.12(b) as a light gray plane—it needs high absolute values for $\mathrm{Re}(\varepsilon_{yy})$ and $\mathrm{Re}(\mu_{zz})$ to be transparent.

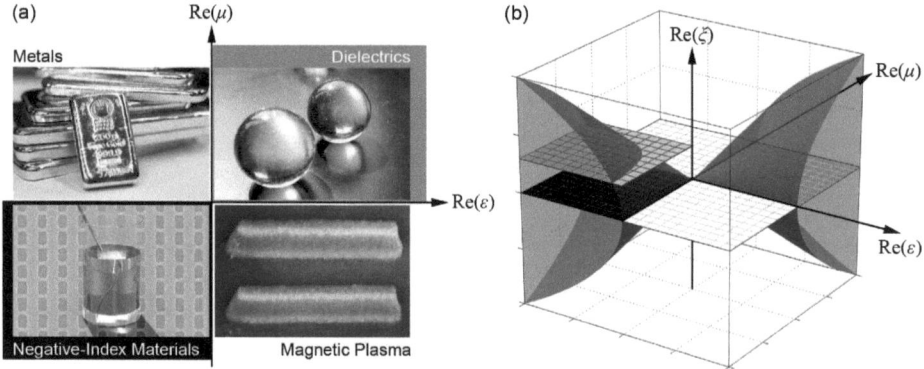

Figure 2.12: Different types of materials categorized by a set of parameters $\mathrm{Re}(\varepsilon_{yy})$, $\mathrm{Re}(\mu_{zz})$, and $\mathrm{Re}(\xi_{yz})$. Note that light propagation occurs only for $n_{\mathrm{ba}}^2 = \varepsilon_{yy}\mu_{zz} - \xi_{yz}^2 > 0$ (see also (2.54)). The shadings for both figure parts are identical, i.e., the parameter region of negative-index materials is shown in black, whereas the parameter region for dielectrics is gray. (a) Categorization of the isotropic class of materials—i.e., $\mathrm{Re}(\xi_{yz})=0$ (q.v. Tab. 2.1). The lower-left and the lower-right picture are taken from Ref. [85] and Ref. [58], respectively. (b) Analog categorization of purely bi-anisotropic materials. Due to cross-coupling effects an additional $\mathrm{Re}(\xi_{yz})$-axis must be introduced. For $\mathrm{Re}(\xi_{yz})=0$, we find again the representation in (a), where two quadrants represent regimes of light propagation though yet projected. For $\mathrm{Re}(\xi_{yz}) \neq 0$, the square-root dependence of the refractive index (2.54) decreases the range of light propagation (light gray plane) which is represented by the darker volume fraction.

For the experimental characterization of bi-anisotropic structures, the respective Fresnel equations are relevant as they link the observable transmittance and reflectance to material parameters [37,82,83]. A detailed derivation in section A.5 [51,84] yields the impedance as a function of the reflection coefficients $r_\pm$, the transmission coefficients $t_\pm$, and the relative impedances $z_{1,2} = Z_{1,2}/Z_0$ of the background media for each propagation direction indicated by the algebraic signs ($\pm$) (see Fig. A.2) [32]:

$$Z_\pm = Z_0 \left(\frac{-b \mp \sqrt{b^2 - 4ac}}{2a} \right) , \qquad (2.55)$$

where a, b, and c are given by

$$\begin{aligned} a &= t_+ t_- - (1 - r_+)(1 - r_-) , \\ b &= (z_1 - z_2)(t_+ t_- + 1 - r_+ r_-) + (z_1 + z_2)(r_+ - r_-) , \\ c &= z_1 z_2 \left[-t_+ t_- + (1 + r_+)(1 + r_-) \right] . \end{aligned}$$

In addition, the implicit expression for the refractive index reads

$$\cos(nk_0 d_\mathrm{s}) = \frac{t_+}{2} \left(\frac{1 - z_-/z_2}{1 + r_+ - (1 - r_+)z_-/z_1} + \frac{1 - z_+/z_2}{1 + r_+ - (1 - r_+)z_+/z_1} \right) , \qquad (2.56)$$

where d_s denotes the slab thickness of the purely bi-anisotropic medium.

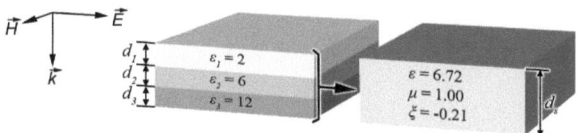

Figure 2.13: Stack of three dielectric materials with (real) dielectric constants $\varepsilon_1=2$, $\varepsilon_2=6$, and $\varepsilon_3=12$ and thicknesses $d_1 = d_2 = d_3 = 10\,\text{nm}$ embedded in vacuum breaks inversion symmetry. For example, at 1 µm wavelength and for normal incidence of light, the physics can be described by a single effective slab ($d_s = 30\,\text{nm}$) with optical parameters $\varepsilon = 6.72$, $\mu = 1.00$, and $\xi = -0.21$. Taken from Ref. [51].

A closer look at the derivation reveals that the transmittance through a slab is identical for propagation in $+x$- and $-x$-direction, i.e, $T_+ = T_-$. However, the complex transmission coefficients $t_\pm$ might have different phases. In contrast, neither the reflectance $R_\pm$ nor the absorbance $A_\pm = 1 - T_\pm - R_\pm$ is generally symmetric.

For illustration, let us consider the simplified example shown in Fig. 2.13. The three dielectric layers can be viewed as one ($N=1$) unit cell of a periodic structure that has no centrosymmetry along the propagation direction of light. Performing a bi-anisotropic retrieval on this configuration at, e.g., $\lambda = 1\,\mu\text{m}$ wavelength ($\lambda \gg d_1 = d_2 = d_3 = 10\,\text{nm}$) leads to the effective material parameters $\varepsilon = 6.72$, $\mu = 1.00$, and $\xi = -0.21$. The layers refer to a fictitious single homogeneous effective slab with total thickness $d_s = d_1 + d_2 + d_3 = 30\,\text{nm}$. We have explicitly verified that the same parameters are retrieved if $N = 2, 3, 4, \ldots, 20$ unit cells of the identical three-layer structure are considered (i.e., the total slab thickness is $N \times 30\,\text{nm}$). Thus, the retrieved quantities can indeed be interpreted as effective material parameters. As the damping is strictly zero for dielectrics, no absorption occurs. Hence, the sum of transmittance and reflectance is unity for each propagation direction. Both reflectances $R_\pm$ are identical in this case and differences only occur in the phases of the complex reflection coefficients $r_\pm$.

This example clearly shows that one should be cautious with using the original Maxwell–Garnett approximation [86] at this point[6], as it would cast the effective behavior of the three sub-wavelength dielectric layers in Fig. 2.13 into just one effective dielectric function, whereas $\mu = 1$ and $\xi = 0$. Hence, the Maxwell–Garnett approximation obviously ignores the direction dependence of the impedance.

2.8.2. Chirality

Another special case of (2.11)–(2.12) arises if induced and exciting field components include an angle of $\varphi = 0°$. This assumption implies that the material parameter tensors $\underline{\varepsilon}$, $\underline{\mu}$, and $\underline{\xi}$ have only diagonal entries. If the respective entries are equal (e.g., $\varepsilon_{xx} = \varepsilon_{yy} = \varepsilon_{zz}$), the parameters can be reduced to complex numbers. Corresponding media are referred to as "bi-isotropic" or

---
[6]Note that extended models of the Maxwell–Garnett effective medium theory [87] can take the bi-anisotropic nature of composites into account.

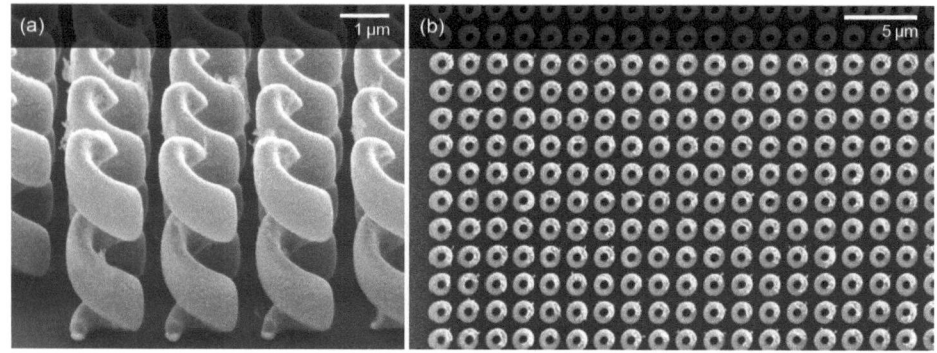

Figure 2.14: Electron micrograph of a left-handed gold helix structure which can be used as a broadband circular polarizer. **(a)** Oblique-view image. **(b)** Top-view image revealing the circular cross section of the helices and the homogeneity on a larger scale. The lattice constant of the square lattice is 2 µm. Both images reproduced with permission from Ref. [33].

"chiral"[7] and obey the constitutive relations given by [51, 81]

$$\vec{D} = \varepsilon_0 \varepsilon \vec{E} - \frac{i}{c_0} \xi \vec{H} \; , \tag{2.57}$$

$$\vec{B} = \mu_0 \mu \vec{H} + \frac{i}{c_0} \xi \vec{E} \; . \tag{2.58}$$

Let us consider a linearly polarized plane wave impinging on a chiral material slab. Here, the incident electric field induces both an electric and a magnetic dipole moment which are oriented either parallel or anti-parallel ($\vec{E} \parallel \vec{P}, \vec{M}$). An analog behavior is also obtained for the magnetic component of the light field ($\vec{H} \parallel \vec{M}, \vec{P}$). As the exciting field components $\vec{E}$ and $\vec{H}$ are perpendicular to each other, vector addition will eventually lead to an effective rotation of $\vec{M}$ and $\vec{P}$. Hence, the initial linear polarization of the incident light is successively rotated as it propagates through the chiral medium. The polarization eigenstates of light are, thus, no longer linear but circular. The emerging optical activity and circular dichroism are related to the real and imaginary parts of the material parameters, respectively.

The associated refractive index of a chiral medium is given by

$$n_{\text{ch}} = \sqrt{\varepsilon \mu} \mp \xi \; , \tag{2.59}$$

where each algebraic sign stands for left- and right-handed circular polarization. Again, the sign of the complex root must be chosen such that $\text{Im}(n) > 0$. In contrast to the pure bi-anisotropic case, (2.59) sustains $n_{\text{ch}} < 0$ for one handedness of light even if both ε and μ are mainly real and positive. For this purpose, only the absolute value of ξ must be large enough. The derivation of the Fresnel equations for chiral metamaterials can be found in Ref. [88].

During the last few years, much effort has been spent to realize artificial chiral materials (dielectric photonic crystals [89–92] as well as metallic metamaterials [93–95]) providing much

---

[7]An object or a system is *chiral* if it cannot be superposed on its mirror image. Chirality (Greek, χειρ: *hand*) or "handedness" is an inherently three-dimensional (3D) phenomenon and occurs, e.g., for DNA, cholesteric liquid crystals, and helical metal antennas.

larger optical activity and circular dichroism than obtained from natural substances (e.g., milk and sugar solution). Recently, a chiral 3D metamaterial has been reported which could be used as a broadband circular polarizer (shown in Fig. 2.14) [33]. To date, similar "classical" polarizers consist of multiple layers of precisely-cut birefringent crystals which are wrung together in optical contact. Hence, fabrication is quite demanding. In contrast, a metamaterial circular polarizer based on an array of gold helices could serve as a fairly simple alternative.

3. Three-Dimensional Metamaterials for Photonics

The discussions in the prior chapter revealed that a variety of novel optical phenomena can be expected once a metamaterial structure is at hand. From an experimental point of view, the fabrication of composite materials for long wavelengths is certainly much easier than for the infrared (IR) or visible (VIS) spectral range. This is mainly due to the scalability of Maxwell's equations and the demand to fulfill (2.2). If the metamaterial should show a magnetic response in the microwave regime from 1 mm to 1 m wavelength, we could use millimeter-sized building blocks [12–14, 21, 96] which can be realized *via* standard printed-circuit-board technology. This technique enables arbitrarily complex structure designs with comparatively low endeavor.

However, for many applications in the field of information technology and optics, the interest is rather focused on the full control of material dispersion in the IR / VIS, i.e., wavelengths between $0.5\,\mu m$ ($\nu = 600\,\mathrm{THz}$) and $15\,\mu m$ ($\nu = 20\,\mathrm{THz}$). Since the periodicity of the metamaterial unit cells still has to be smaller than the wavelength of interest, the required feature sizes push nanofabrication technology to its limits. To get an impression of the targeted feature sizes, we simply relate to the quasi-effective medium limit (2.2) while assuming glass as a background medium with ($n_\mathrm{bg} = n_\mathrm{glass} = 1.5$). Thus, the size of unit cells must be right below $8\,\mu m$. Note that the latter assumption is based on the fact that most photonic metamaterial structures presented to date are *not* self-supporting and, thus, placed on top of a transparent substrate. In the IR / VIS, silica is often used due to its high optical quality while being highly inert. Anyway, due to the stronger condition (2.1), one simple rule always holds: the smaller the better.

Besides the required feature sizes, one should also keep in mind that we are concerned with composites consisting of "natural" materials whose properties strongly depend on the incident light's wavelength. This does not only hold for the permittivity (*q.v.* Drude model in section 2.6.2) but also for other relevant quantities[1]. Thus, simply down-scaling metamaterial structures used in the microwave regime does not necessarily yield meaningful designs at shorter wavelengths [97]. Some additional adaptations are required, too.

Most well-established fabrication techniques for nanoscaled structures involve inherently two-dimensional (2D) "top-down" processes. Indeed, by using state-of-the-art lithographic devices, it is possible to produce *planar* nanometer-sized magnetic building blocks which fulfill the effective medium limit even for the VIS [80, 98]. But using the standard nanofabrication technology yields meta*films* rather than bulk meta*materials*. For such planar structures, the incident light "feels" only one layer of unit cells while propagating perpendicular to the substrate plane. This

---
[1] By way of example, the conductivity of noble metals in the IR / VIS is much lower than for microwaves. The lower conductivity translates to a comparatively high transmittance but also to higher losses due to the increased ohmic resistance.

is definitely contrary to typical model systems in solid-state physics, where multiple lattice constants along all spatial directions are assumed.

3.1. Bulk or Non-Bulk: That's the Question.

Reverting to the previously introduced condition for quasi-effective media (2.2), one might wonder why the realization of bulk metamaterials should be of any importance. Indeed, at first glance, there seems to be no such requirement, since planar metafilms can be treated as optical black boxes as well. But catching a glimpse of the metafilm black boxes' interior clearly reveals that the observed material properties are mainly affected by surface effects.

From research topics in condensed-matter physics, we know that analog configurations of ultrathin material slabs exhibit totally different properties compared to their bulk version. Graphene [99, 100], being a monolayer of carbon atoms tightly packed into a 2D honeycomb lattice, is a very popular example. Transport measurements have demonstrated that graphene has an exceptionally high electron mobility which makes it the best conducting substance known at room temperature. Note that this property can be exclusively attributed to its 2D geometry [101].

In the same way, we anticipate a change in the optical properties of composite materials once converging the bulk limit, where the phase of a light wave is considerably modified while passing through a large amount of unit cells. Here, two pivotal questions arise: (i) How many unit cell layers have to be stacked until we can call a material system "bulk"? (ii) Do the material properties change in a favored or unfavored manner when approaching the bulk limit? Unfortunately, there do not exist general answers to these questions. The bulk limit and the resulting properties must be identified individually for each and every case as the physics strongly depends on the respective configuration.

On the basis of experiments on bulk composites for the microwave regime and "classic" solid-state optics, a pronounced magnetic response is expected once surface effects are negligible. Thus, the motivation to fabricate bulk metamaterials is clearly comprehensible and, in any case, an important issue to gain a deeper insight into the fundamentals of photonic magnetic materials.

Before we see about the fabrication of metamaterials, it is important to point out the differences between *three-dimensional* and *bulk* media, since there might appear confusions regarding the definitions: In this Thesis, all structures consisting of at least one layer of non-planar unit cells or a few layers of planar unit cells will be called "three-dimensional" metamaterials. They form the intersection between planar and bulk materials with surface effects still being very pronounced. Stacking many layers of 3D structures results in materials whose optical behavior is increasingly dominated by propagation effects. Only if adding an additional layer of unit cells does not alter the material parameters anymore, the structure will be called "bulk". Hence, realizing 3D composite structures is an intermediate but important step towards bulk metamaterials.

3.2. Layer-By-Layer Approaches Towards Three-Dimensional Metamaterials

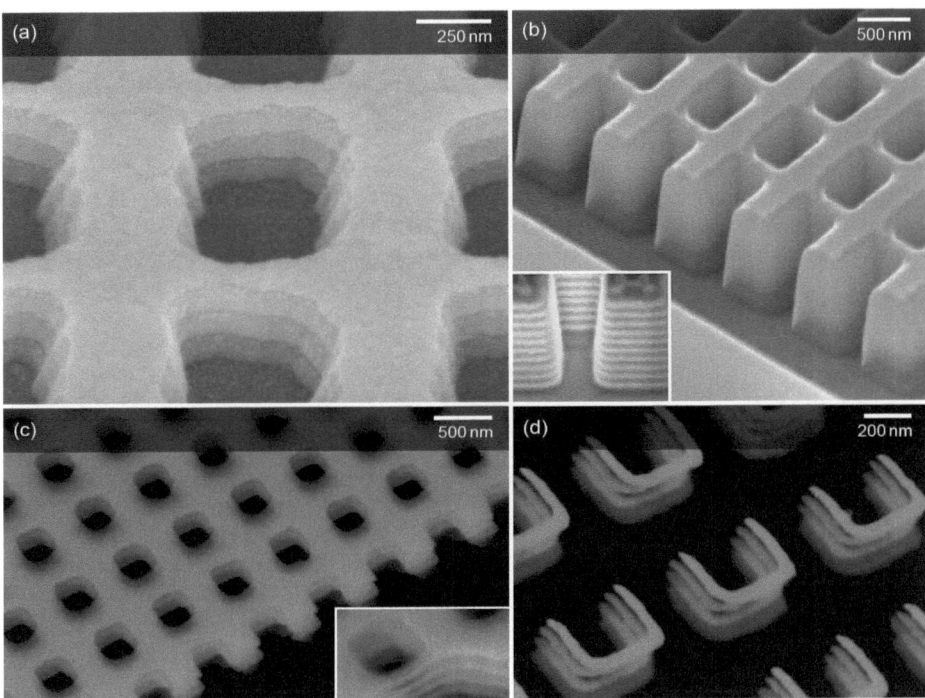

Figure 3.1: Electron micrographs of metamaterials fabricated *via* layer-by-layer techniques. (a) Three layers of a fishnet structure fabricated by electron-beam lithography (EBL) and subsequent lift-off process. Taken from Ref. [59]. (b) Ten functional layers of a fishnet structure fabricated by focused-ion beam milling of multiple dielectric (dark) and metallic (light) layers. The inset clearly reveals the tapering effect. Reproduced with permission from Ref. [31]. (c) Five layers of a fishnet structure fabricated by EBL and planarization. Reproduced with permission from Ref. [29]. (d) Four layers of split-ring resonators realized by EBL and planarization. Reproduced with permission from Ref. [28].

First attempts to fabricate photonic metamaterials used standard lithographic techniques such as electron-beam lithography (EBL) [24–27], focused-ion-beam (FIB) lithography [31, 102], and ink-jet printing [103, 104] which are all established nanotechnologies, mostly used for maskless lithography of low-volume production by semiconductor industry as well as research and development. Anyway, these tools were not meant to be used for realizing 3D composites. Hence, during the last years, the metamaterial community spent a lot of effort to extend the aforementioned techniques for 3D processing and, indeed, found some remarkable solutions. Regrettably, all of them—to be discussed in the following sections 3.2.1 and 3.2.2—have certain geometrical constraints which restrict their application to distinct unit cell designs.

3.2.1. Single-Step Structuring

Serial evaporation of multiple alternating metallic and dielectric layers and subsequent structuring makes it possible to stack many functional layers on top of each other. The structuring step is accomplished either *via* focused-ion beam (FIB) milling [31], i.e., sputtering by using gallium ions, or by a standard lift-off process [30, 59].

Metamaterials which are fabricated by means of single-step structuring are shown in Fig. 3.1(a) and (b). Surveying the resulting structures reveals that one has to deal with significant tapering effects which break the symmetry along the light path. Notably, this gives rise to bi-anisotropy—which has been often neglected in literature—and might deteriorate the optical properties, i.e., a broadening of magnetic resonances and an increase of damping [105]. Beyond that, the maximum number of processable layers is fairly limited: Three and ten functional layers of a fish-net structure have been realized by using a standard lift-off process [30] and FIB milling [31], respectively. Note that these demonstrations are, to date, the upper limit of technical feasibility.

3.2.2. Planarization

Another layer-by-layer fabrication approach uses EBL in combination with fluid spin-on dielectrics [106]. The whole processing cycle is illustrated in Fig. 3.2. At first, a single layer of metallic unit cells is produced by lithography (Fig. 3.2(1–5)). Afterwards, a fluid dielectric is spun-on the structure (shown as a dark film in Fig. 3.2(6–8)) which planarizes the metallic layer and, in addition, isolates consecutive metal layers electrically. After curing, another metal layer is stacked on top (Fig. 3.2(8)). The latter step is a non-trivial procedure since an alignment with nanometer-precision is required.

The striking advantage of this method is its flexibility to design arbitrary planar metallic features in lateral directions parallel to the substrate plane [28, 29, 107, 108]. Some examples are presented in Fig. 3.1(c) and (d). The main drawback of the planarization technique is, certainly, the serial processing. Since inherent 2D techniques are used, each step must be repeated many times to realize 3D structures (... not to mention the bulk composites) which makes the fabrication very time-consuming.

3.3. Inherent Three-Dimensional Fabrication

2D stacking techniques are, apparently, very important for first analyses of 3D metamaterials, since they enable to transfer results of planar structures to the 3D world—in close analogy to single graphene layers which are stacked on top of each other to yield bulk graphite. In other words, we can start from a simplified system (e.g., a single SRR) and extend the well-understood model to our needs. However, for realizing bulk or even isotropic 3D metamaterials, a truly 3D fabrication approach would be preferable. Hence, we take a step forward and ask for alternatives by replacing the inherently 2D processes by their 3D analogues. As we require additional degrees of freedom for each fabrication step, a 3D lithography method which allows for structuring of photoresists in all three dimensions as well as appropriate metallization processes are sought.

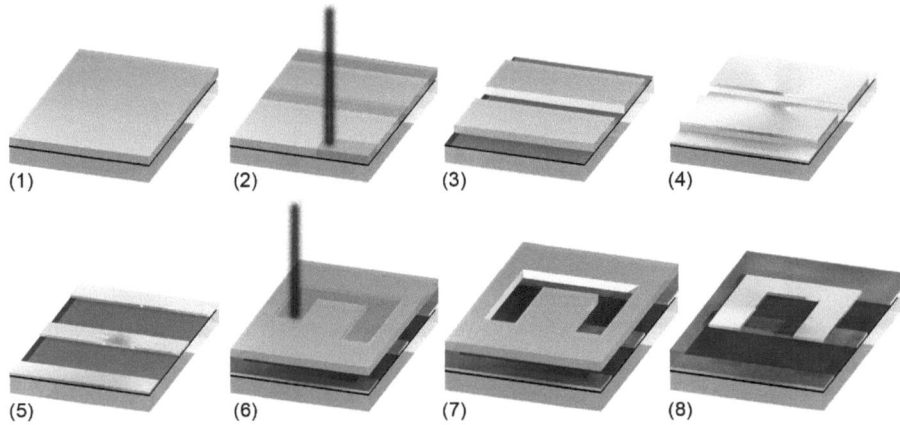

Figure 3.2: Illustration of the layer-by-layer fabrication process which utilizes an intermediate planarization step. The production sequence is indicated by numbers in ascending order. (**1**) Fabrication starts with a glass substrate coated with a thin film of indium tin oxide (ITO) and a spun-on photoresist like, e.g., poly(methyl-methacrylate) (PMMA). ITO is an optical transparent semiconductor which has a sufficient d.c. conductivity to prevent electric charging during the EBL process. (**2**) The photoresist has to be sensitive to electron beams (dithering lines) such that it changes its chemical properties and becomes soluble for appropriate developers. (**3**) During the development step, the exposed (in case of a positive-tone resist) or the unexposed (in case of a negative-tone resist) volumes are removed. (**4**) By using a highly-directional physical evaporation technique (e.g., electron-beam evaporation or plasma sputtering), the whole substrate is metallized (shiny). (**5**) The subsequent lift-off process ends the fabrication of the first functional layer. (**6**) A "spin-on dielectric" (dark gray) and another film of photoresist are spun-on the first layer. After exposure, (**7**) development, metallization, and lift-off process, (**8**) the second functional layer is finished.

It is possible to rapidly realize large-scale 3D polymer templates by using holography (interference lithography) or self-assembly of colloidal particles. However, these techniques do not exhibit enough degrees of freedom to fabricate arbitrarily shaped structures. We will see in chapter 5 that it is fairly hard to think of simple-looking designs for bulk magnetic metamaterials. Hence, for cutting-edge developments, we need, first of all, as much flexibility as possible and do not care that much about mass production.

In short terms, our basic idea [32, 109–113] is to use 3D direct laser writing (DLW) [114–116] to fabricate polymer templates (section 3.3.1) which are subsequently protected by a thin inert oxide ceramic layer (section 3.3.2) [117–119] and, finally, metallized (section 3.3.3) *via* chemical vapor deposition (CVD) [120, 121] or electroplating [33]. Besides metallization of polymer templates, there also exist first attempts to write 3D structures directly into a metal-containing resist, i.e., silver nitrate ($AgNO_3$) in aqueous solution combined with a coumarin 440 ethanol solution [122, 123]. When illuminated by a focused laser beam, metal-ions in the photoresist absorb two photons simultaneously and reduce to pure metal. Because of the emerging silver features' high reflectivity, the laser beam is deflected during the writing process. Hence, the structure quality is highly deteriorated. This manifests itself in large and fairly rough metal features (on the micrometer scale) which is a no-go criterion for fabricating 3D photonic metamaterials.

3.3.1. Three-Dimensional Polymer Templates by Direct Laser Writing

DLW (also known as direct laser lithography) of polymer templates has been known since years by the photonic crystal community [8, 124]. Similar to standard photolithography techniques, structuring is accomplished by illuminating negative-tone or positive-tone photoresists *via* light of a well-defined wavelength. The fundamental difference is, however, the avoidance of reticles. Instead, two-photon absorption[2] is utilized to induce a dramatic change in the solubility of the resist for appropriate developers. Since two-photon absorption is a second-order, non-linear process several orders of magnitude weaker than linear absorption, very high light intensities are required to increase the number of such rare events. For example, tightly-focused laser beams provide the needed intensities. Here, pulsed laser sources are preferred as they deliver high-intensity pulses while depositing a relatively low average energy[3].

To enable 3D structuring, the light source must be adequately adapted to the photoresist in that single-photon absorption is highly suppressed while two-photon absorption being favored. This condition is met if and only if the resist is highly transparent for the laser light's output wavelength λ_out and, simultaneously, absorbing at $\lambda_\text{out}/2$. As a result, we can scan a given sample relative to the focused laser beam while changing the resist's solubility only in a confined volume. The geometry of the latter mainly depends on the iso-intensity surfaces of the focus (see Fig. 3.3(b)). Concretely, those regions of the laser beam which exceed a given exposure threshold of the photosensitive medium define the DLW's basic building block, the so-called "voxel". Other parameters which influence the actual shape of the voxel are the laser mode and the refractive-index mismatch between the resist and the immersion system leading to spherical aberration.

For all dielectric templates presented in chapter 4, we used the commercially available negative-tone photoresist SU-8 (epoxy resin by MicroChem Corp.). It has on average eight epoxy groups per monomer and contains a triaryl sulfonium salt as cationic photoacid generator [124]. When the photoresist is illuminated, the included photoacid generator breaks the bonds of the monomers and acts as a catalyst for a spatially-defined chain-growth polymerization. Generally, SU-8 requires additional processing steps, such as a pre- and post-exposure bake, to remove the solvent and accelerate cross-linking, respectively. Subsequent developing by using isopropyl alcohol or ethyl lactate releases the non-polymerized resist and leaves behind a mechanically stable 3D polymer template like shown in Fig. 3.4.

---

[2]Two-photon absorption [125] is based on the simultaneous absorption of two photons of identical (or different) frequencies in order to excite a molecule from an initial to a higher energy electronic state. The related energy difference between the involved states is equal to the sum of the energies of both photons ($\Delta E = \hbar\omega_1 + \hbar\omega_2$). It differs from linear absorption in that the strength of absorption depends on the *square* of the light intensity which makes it a non-linear optical process.

[3]Light absorption produces heat. Hence, if the deposited average energy is too high, the photoresist will be burned.

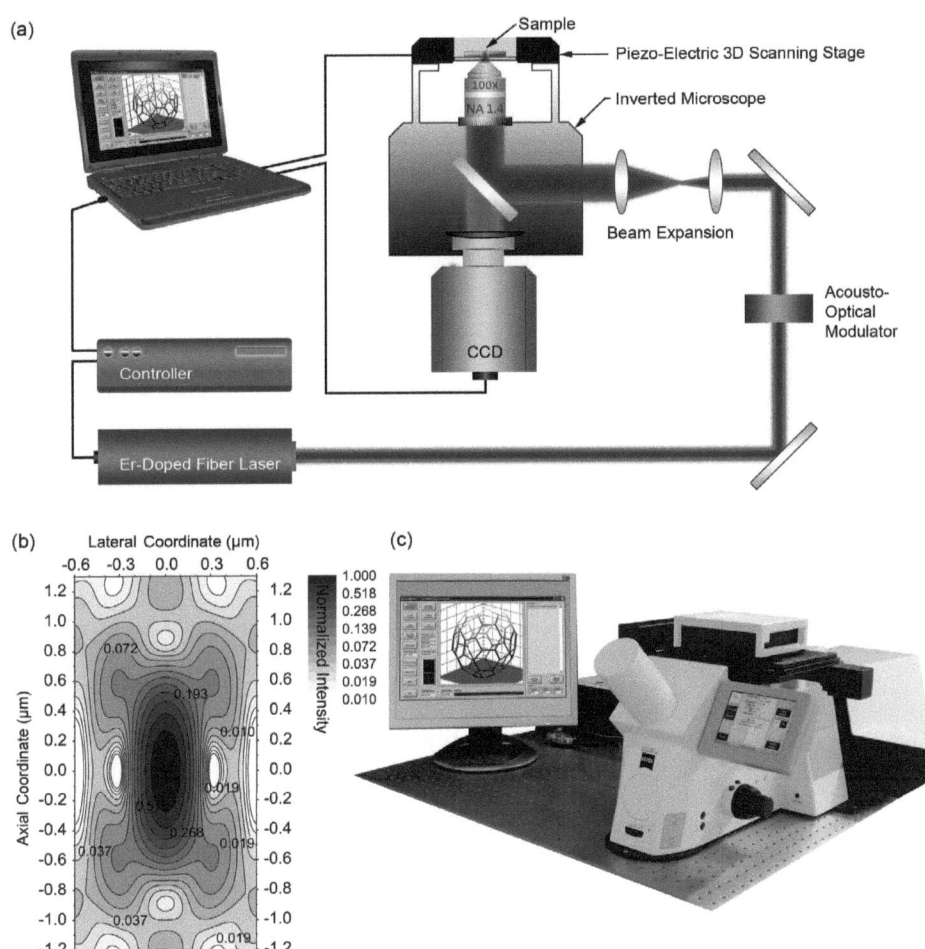

Figure 3.3: Principle of direct laser writing (DLW). (a) Scheme of the DLW setup used to fabricate polymer templates for 3D photonic metamaterials. A compact frequency-doubled Erbium-doped fiber laser at 100 MHz repetition rate and sub-150 fs pulses is used at a central wavelength of 780 nm. The laser power (approx. 60 mW) is attenuated by using an acousto-optical modulator. The 3D scanning-piezo stage (scanning volume of 300 μm × 300 μm × 80 μm), the CCD camera, and the laser system are controlled *via* PC interface. Courtesy of Nanoscribe GmbH. (b) Iso-intensity map of a focused laser beam plotted over axial and lateral coordinates. If the black inner area corresponds to the threshold energy needed to polymerize a photoresist, one obtains an ellipsoidal voxel. (c) Photograph of Nanoscribe's "Photonic Professional" DLW system. Courtesy of Nanoscribe GmbH.

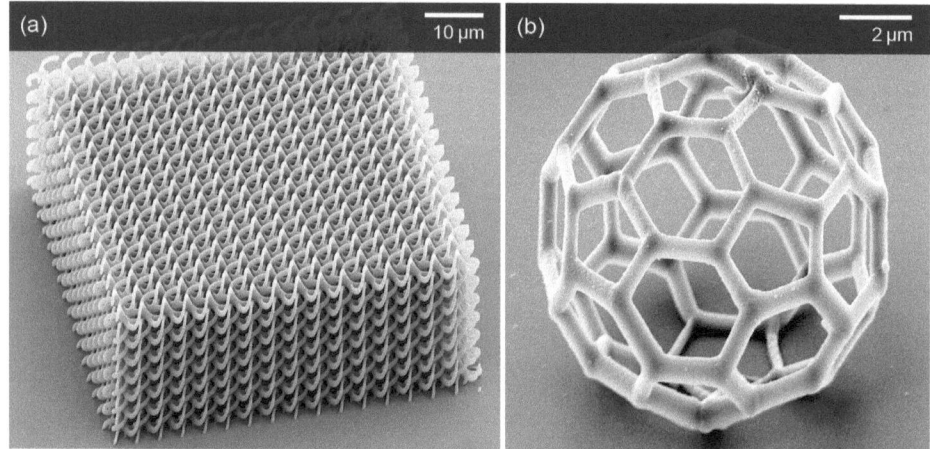

Figure 3.4: Examples of dielectric structures both of which are fabricated by using Nanoscribe's direct laser writing system (Photonic Professional, see Fig. 3.3(c)). **(a)** Bi-chiral photonic crystal. Taken from Ref. [92]. **(b)** Polymeric bucky ball as an example of an open but still mechanically stable 3D structure. Courtesy of Nanoscribe GmbH.

Alternatively, also positive-tone photoresists like AZ 9260 (by MicroChemicals GmbH) are available. In this case, only those regions which are sufficiently exposed by light are removed by the developer (potassium-hydroxide-based inorganic solution).

3.3.2. Protection of Polymers by Highly Stable Oxide Ceramics

To prevent the templates from melting and unintended chemical reactions during post-processing, a thin stabilizing dielectric layer must be coated on top of the polymer surface. Of course, this layer has to be applied by using a compatible process which does not alter or deteriorate the original structure. We found pulsed layer deposition (PLD) of silica (SiO_2) and atomic layer deposition (ALD) of titania (TiO_2) to be useful for our purposes.

Pulsed layer deposition of silica: The deposition of silica is performed at room temperature and atmospheric pressure in a gas-tight glass reactor which is connected to reservoirs of silicon tetrachloride ($SiCl_4$) and water (H_2O) *via* computer-controlled valves. To induce the chemical reaction [117]

$$SiCl_4 + 2\,H_2O \rightarrow SiO_2 + 4\,HCl\;,$$

H_2O is introduced to the reaction chamber in gaseous phase by an inert carrier gas, e.g., nitrogen (N_2). After a thin layer of H_2O is adsorbed on the template surface, gaseous $SiCl_4$ is passed into the chamber which locally reacts with the already present H_2O. As a consequence, a thin silica layer (typically 3 nm thick) is created on top of the surface. The resulting HCl is carried out of the chamber by using a carrier gas. The sequence of the whole process is illustrated in

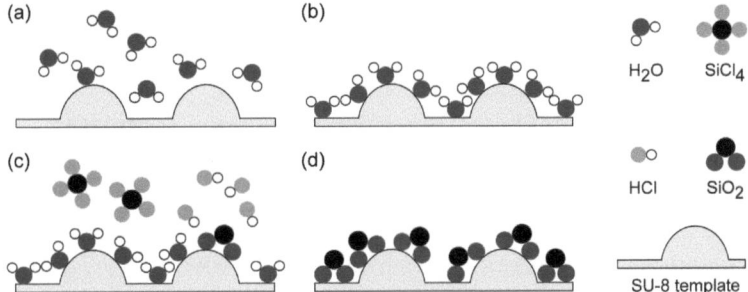

Figure 3.5: Illustration of the pulsed layer deposition process to coat a polymer (SU-8) template with silica. **(a)** Water (H_2O) is introduced in gaseous phase by a carrier gas. **(b)** A thin layer of H_2O is adsorbed on the whole surface of the template. **(c)** Silicon tetracholride ($SiCl_4$) is introduced in gaseous phase and locally reacts with H_2O. **(d)** A thin silica (SiO_2) layer is coated on top of the whole template surface while co-producing hydrochloric acid (HCl). The latter is then transported out of the reaction chamber by using a carrier gas. Adapted from Ref. [84].

Fig. 3.5. A cyclic repetition of this procedure results in closed, robust silica films. In analogy to conventional glass, the deposited silica has a refractive index of $n_{\text{silica}} = 1.45$ at 1 µm wavelength.

Atomic layer deposition of titania: The deposition of titania protection layers works similar to silica. In Fig. 3.5, $SiCl_4$ just has to be replaced by $TiCl_4$. Additional heating is required during step (c) to induce the chemical reaction [118, 119]

$$TiCl_4 + 2\,H_2O \rightarrow TiO_2 + 4\,HCl\,.$$

The process is performed under high vacuum at a temperature of 110°C. Notably, the polymer SU-8 is mechanically stable up to about 120°C. Since the deposited titania is amorphous, the refractive index can be taken as $n_{\text{titania}} = 2.05$ at 2 µm wavelength (value extrapolated from Ref. [126]).

The main difference between both coating processes lies in the control of thickness and the required equipment. While in the case of titania an *atomically* thin layer is created during each cycle, the silica PLD deposits several nanometers at once[4]. Thus, by using an ALD process, one can adjust the target thickness in a very controlled fashion. Moreover, from our experience, the titania surfaces are advantageous compared to silica for the growth of smooth thin metal films (*q.v.* section 3.3.3) which can be accredited to its higher surface energy. However, the titania ALD needs more advanced vacuum and heating equipment.

For the sake of avoiding confusions, it should be mentioned that a self-made silica PLD [127] was already established at the very beginning of our investigations. Thus, some of the polymer templates presented in this Thesis have been protected by silica. After a commercial ALD

---

[4] Actually, this is the reason why both techniques have different names.

(Savannah S-100 by Cambridge NanoTech, Inc.) became available, we switched to titania due to the aforementioned advantages[5].

3.3.3. Metallization of Three-Dimensional Polymer Templates

In the prior chapter, we have noticed that metals are crucial in the context of metamaterials since only conductive features allow for LCR resonances. Thus, it is important to find suitable techniques to metallize 3D templates. Although sputtering and electron-beam evaporation yield very smooth and homogeneous films, only those parts of a template are coated which are placed directly in sight of the metal target. Consequently, the inner surfaces of a 3D template would be useless as they cannot contribute to the required magnetic resonances. Beyond that, we discussed earlier that noble metals like silver (Ag) and gold (Au) occupy the lowest damping factors (q.v. Tab. 2.2) which is directly related to lower losses of the incident light.

Summing up, fully isotropic coating techniques of noble metals are essential. In the following context, two complementary approaches are presented which can be either used in combination with negative-tone or positive-tone photoresists, i.e., CVD of silver [32,109] and electroplating of gold [33]. There also exist some other isotropic metallization methods like electroless deposition [128–131] and metal ALD [132–134], although thin films of sufficiently high optical quality have been reported only recently [135].

Chemical vapor deposition of silver: The CVD of silver [120] is based on a heat-induced decomposition of the ligand-stabilized silver β-dikentonate precursor (1,5-cyclooctadiene)(1,1,1,5,5,5-hexafluoro-acetylacetonato)silver$^{(1)}$ ((COD)(hfac)Ag), to pure silver, metal-organic, and organic byproducts [121]. The chemical reaction is given by

$$(\text{COD})(\text{hfac})\text{Ag}_{(\text{gaseous})} \xrightarrow{\text{adsorption}} (\text{hfac})\text{Ag}_{(\text{adsorbed})} + (\text{COD})_{(\text{gaseous})} \qquad (3.1)$$

$$2\,(\text{hfac})\text{Ag}_{(\text{adsorbed})} \xrightarrow{\text{heating}} \text{Ag}_{(\text{solid})} + \text{Ag}(\text{hfac})_{2\,(\text{gaseous})}\,. \qquad (3.2)$$

Analog to the already presented titania ALD, we use a cyclic process. Each CVD cycle begins with the sublimation (3.1) of the solid metal-organic precursor at 65°C at low pressure (< 0.5 mbar) which fills the whole reaction chamber with (COD)(hfac)Ag due to gas diffusion. The sublimation temperature must not be chosen too high as, otherwise, the metal-organic substance decomposes mainly in the crucible before being adsorbed at the substrate surface.

After approximately 45 minutes, enough precursor molecules are adsorbed to initiate the thermal decomposition (3.2). Here, the substrate temperature influences the kinetics of the chemical reaction. If the temperature is below 150°C, the metal film predominantly suffers from organic contamination. If it is higher than 170°C, the surface mobility of the silver atoms increases such that the films tend to become more granular (q.v. Fig. 3.6(c)) which again deteriorates the optical properties. From our experience, 160°C–165°C are the decomposition temperatures giving the best compromise. The decomposition step takes 10 minutes and results in the deposition of a 3 nm-thick silver film. To further improve the quality of the coatings, we activate the

---

[5]We do not expect an appreciable modification of the metamaterial's optical properties caused by the change of the protecting layer. Note that the change from $\varepsilon_{\text{silica}} = n_{\text{silica}}^2$ to $\varepsilon_{\text{titania}} = n_{\text{titania}}^2$ is relatively small compared to the permittivity of metal (to be added in the next fabrication step).

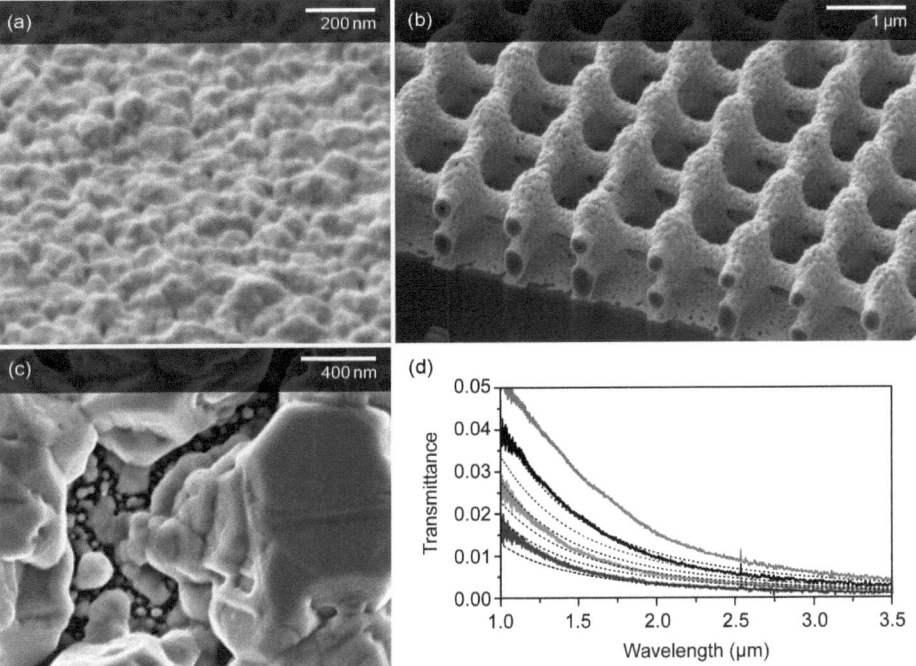

Figure 3.6: Electron micrographs of silver films deposited *via* chemical vapor deposition of (COD)(hfac)Ag. **(a)** The silver film on a plane glass substrate looks somehow rough, but the grain size (some tens of nanometers in diameter) is still small compared to the wavelength of the incident light (in the range of few microns). Importantly, the grains of the metal layer are electrically connected. **(b)** In contrast to sputtering techniques, chemical vapor deposition delivers spatially isotropic silver films (light regions). The depicted test structure is cut by using focused-ion beam milling to reveal that the wires are coated all around. Taken from Ref. [32]. **(c)** Using an extraordinarily high deposition temperature (here 190°C) results in large metal clusters. This is mainly caused by the high mobility of silver on the glass substrate at elevated temperatures. **(d)** Optical properties of closed plane silver films like shown in (a) and (b) can be nicely fitted by the Drude free electron model. The gray shaded curves indicate the measured spectra. The measurement was performed on metallized plane samples (not shown) by using a Fourier-transform infrared-spectrometer (*q.v.* section 3.4.2). The dotted black lines correspond to calculated spectra using a scattering-matrix approach (see section 3.4.1) and the Drude parameters for silver from Tab. 3.1. The calculated layers have nominal thicknesses of 36 nm, 40 nm, 42 nm, 44 nm, 46 nm, and 48 nm, whereas thinner metal layers exhibit a higher transmittance.

dielectric surface of the template by applying air plasma (PlasmaPrep5 by Gala Instrumente GmbH) for 15 minutes to the sample before starting the metallization. Moreover, the metal-organic precursor, being sensitive to oxygen, must be kept under nitrogen atmosphere as long as possible before evacuation.

During the last step, the reactor chamber is evacuated. While silver remains on top of the template, the metal-organic byproducts are pumped out of the chamber. Repeating this cycle 10 to 15 times results in an electrically connected metal coating with a thickness of about 40 nm (see Fig. 3.6(a)). During the whole process, the walls of the reaction chamber are heated to 110°C to avoid condensation of the precursor.

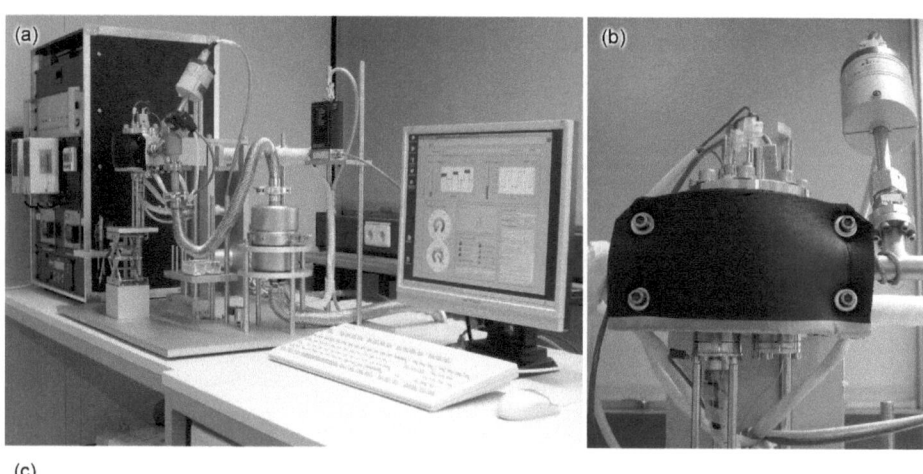

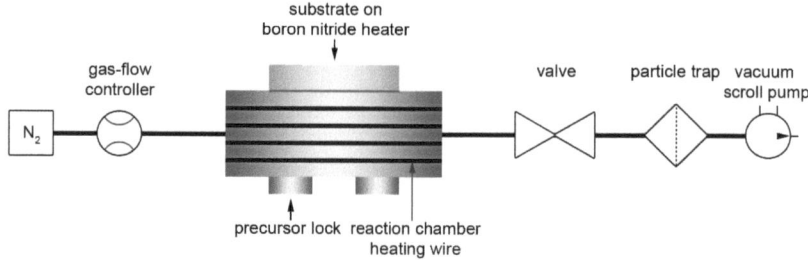

Figure 3.7: Self-made chemical vapor deposition (CVD) system used to deposit isotropic silver coatings. (a) Overview of the whole setup. From left to right: (i) Control tower containing all controlling devices, i.e., the power supply, outer-chamber heaters, inner-chamber heaters, and the gas-flow controller. (ii) Reaction chamber, valves and particle trap. (iii) PC for controlling all devices inside the tower (i). The automation of the CVD enables the control of the heating temperatures and a temporal control of the cycles. (b) Detailed view of the reaction chamber showing the precursor lock at the underside and the sample lock at the upper side of the reaction chamber. (c) Schematic of the reactor's line system.

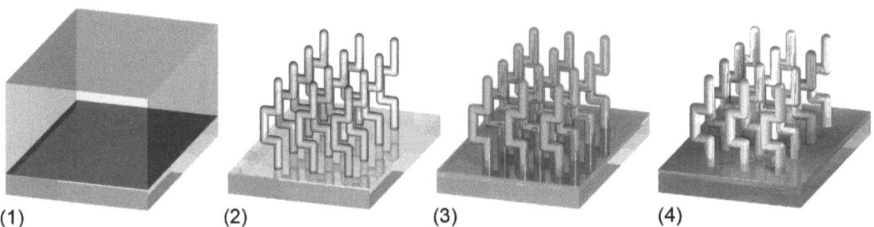

Figure 3.8: Overview of the whole 3D fabrication method *via* direct laser writing (DLW) and chemical vapor deposition (CVD) of silver. **(1)** The sequence starts with spinning a negative-tone photoresist like SU-8 onto a glass substrate. **(2)** Connected 3D polymer templates are fabricated by using DLW and subsequent developing. **(3)** A silica or titania coating is applied to protect the polymer template from melting and unintended chemical reactions during the metal deposition. **(4)** Isotropic metallization (typically several nanometers in thickness) *via* CVD of silver.

As no CVD reactor suiting our specifications has been commercially available, we built our own automated setup shown in Fig. 3.7. The automation *via* Labview programming language (National Instruments Corporation) helps to obtain reproducible results and allows to monitor all relevant experimental parameters, i.e., the substrate / precursor temperature and the chamber pressure. Importantly, all CVD chamber parts have to be made of chemically compatible or highly inert materials (stainless steel, nitride ceramics, poly(tetrafluoroethylene), etc.) in order to prevent reactions with the organic byproducts. Especially, standard copper gaskets should be avoided since there exists a connatural precursor (hfac)(COD)Cu which might be produced inside the chamber during silver deposition. The reactor chamber must be cleaned after each and every CVD process to reduce contamination by remaining organic molecules.

In Fig. 3.6(a) and (b), resulting coatings are shown as electron micrographs. They exhibit a good d.c. conductivity and a reflectance $R > 95\%$ in the near-IR (from 1.0 µm to 4.0 µm). Beyond that, the transmittance spectra fit well to the Drude model (*q.v.* Fig. 3.6(d)) which is an important indicator for a high purity and optical quality. Clearly, the silver films are somehow rough, but the grain size (approximately 20–30 nm in diameter) is small compared to wavelengths at which the metamaterials will show their magnetic resonances (i.e., from 1.0 µm to 4.5 µm). As shown in Fig. 3.6(b) for a cubic-wire template, the silver coating is isotropic. If we had used an anisotropic metallization method, just the upper wires of the framework would have been coated, whereas the whole surface, especially the interior of the structure, is metallized *via* the CVD method. Note that it is not possible to create separated metal parts along a stabilizing wire connection without locally-defined functionalization. This is an essential point that has to be taken into account when designing metamaterial structures. Moreover, the template-sustaining glass substrate is metallized, too, which additionally constitutes a severe constraint (see also Fig. 3.8(4)).

Electroplating of gold: For electroplating of gold (Au)[6] [33], a sulfite-based electrolyte solution is used containing the following ingredients:

(i) sodium disulfitoaurate[(I)] ($Na_3[Au(SO_3)_2]$) as gold precursor

(ii) sodium sulfite (Na_2SO_3) as well as ethylenediamine ($C_2H_4(NH_2)_2$) for chemical stabilization

(iii) ethylenediamine-tetraacetic acid disodium salt dihydrate Na_2EDTA ($C_{10}H_{16}N_2Na_2O_8 \times 2\,H_2O$) to sequester metal ions

The pH value of the solution is adjusted to 8.5. The cathode is a 25 nm-thin ITO layer (sheet resistance of 500 Ω) which is sandwiched between glass substrate and photoresist. The anode is a platinized titanium mesh. During growth, the temperature of the electrolyte is actively stabilized to 57°C using a thermometer coupled to a hot plate. Typically, we use a constant electric current in the order of 10^{-6} A (corresponding to a current density of 10^{-3} A/cm^2). Finally, to remove the polymer, the samples are exposed to air plasma for 24 hours (PlasmaPrep5 by Gala Instrumente GmbH).

In general, if one utilizes CVD for metallization, it is favorably to use a negative-tone photoresist where the polymerized (scanned) structure remains after development (*q.v.* Fig. 3.8(2)). By contrast, one rather uses templates made of a positive-tone photoresist for electroplating. Here, the scanned regions are removed such that the intended structure is represented by holes (*q.v.* Fig. 3.9(2)). Afterwards, the latter are gradually filled with metal from the bottom up (*q.v.* Fig. 3.9(3)) by applying a very defined current to the electrolyte solution. Note that both approaches do not allow for separated metallic features without local surface functionalization of the photoresists. This is a crucial difference to the aforementioned planarization method, where the metallic elements are embedded in a dielectric host material. When thinking of meaningful metamaterial designs, we have to take this constraint into account.

3.4. Analysis of Photonic Metamaterials

The analysis of metamaterials can be either done experimentally or *via* numerical calculations. The overall aim is to obtain a complete set of optical parameters which unambiguously describe the metamaterial's behavior. Remembering the results of the prior chapter, such a set of parameters can be, e.g., the complex electric and magnetic fields—as described by Maxwell's equations (2.13)–(2.16)—or, alternatively, the refractive index n and the impedance Z.

For metamaterials working in the microwave regime, the characterization is again fairly simple. Since the feature sizes are in the range of millimeters, special hardware probes can be placed inside the structure which directly measure the fields at every spatial position. Moreover, typical properties of negative-index materials, such as the beam displacement, can be measured by placing macroscopic detectors around the object. Unfortunately, corresponding methods are not available for nanoscale 3D photonic metamaterials. Scanning near-field microscopy, e.g.,

---

[6]The electroplating process discussed in this section and used to metallize the structures in chapter 5 has been established by Justyna K. Gansel at the Institut für Angewandte Physik (Karlsruhe Institute of Technology).

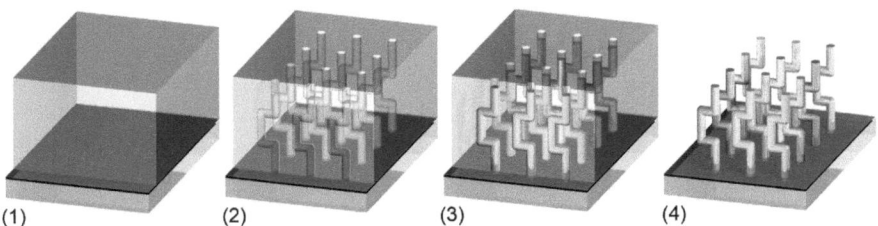

Figure 3.9: Overview of the 3D fabrication procedure which involves direct laser writing (DLW) and electroplating of gold. For clarity, the photoresist's front end is partly removed. **(1)** The sequence starts with spinning a positive-tone photoresist like AZ 9260 onto an ITO-coated glass substrate. The ITO layer serves as a cathode during the electroplating in step (3). **(2)** Connected 3D channels are fabricated by DLW and subsequent developing. Note that the holes must reach down to the ITO layer, otherwise the cathode is electrically isolated during the next step. **(3)** Gradually filling up the holes with gold *via* electroplating results in metallic nanostructures. **(4)** By etching with air plasma for several hours, the remaining photoresist is removed leaving the intended structure.

enables to probe the magnetic fields of planar nanostructures [136]. Due to the outer dimensions of a typical near-field microscope tip (several hundreds of microns) it is, however, not possible to probe the inside of non-open 3D structures with typical feature sizes < 10 µm. Measuring other attributes of photonic metamaterials (e.g., beam displacement of negative-index materials) is also challenging, because most effects happen on a very small scale. Notable exceptions are reported in Refs. [31, 68].

Hence, the easiest approach to obtain a physical understanding of the metamaterial structures uses the comparison of spectral measurements and numerical calculations. Concretely, we simulate the anticipated properties by using adequate calculation tools to be discussed in section 3.4.1. These software packages deliver the electromagnetic fields and, herefrom, the complex transmission and reflection coefficients t and r, respectively. The quantities t and r also serve as a starting point to retrieve optical parameters like n and Z ($q.v.$ the Fresnel equations for isotropic (2.26)–(2.27) and bi-anisotropic (2.55)–(2.56) media). Subsequently, calculations are brought into coincidence with the measured transmittance $T \sim |t|^2$ and reflectance $R = |r|^2$ (for methods see section 3.4.2).

3.4.1. Calculation of Electromagnetic Fields and Optical Spectra

For numerical calculations of metallic structures, we normally utilize the finite-integration technique [137, 138] which is implemented in the commercial software package Microwave Studio (by CST AG). Some related background information about finite-difference time-domain and finite-integration algorithms is given in the appendix A.6.

Microwave Studio supports two different simulation modes, i.e., the amplitudes and phases in both transmittance and reflectance can be calculated either in frequency domain or in time domain. For our purposes, we use the transient solver which operates in the time domain. This means that the evolution of the fields is computed for each discrete time step at discrete points

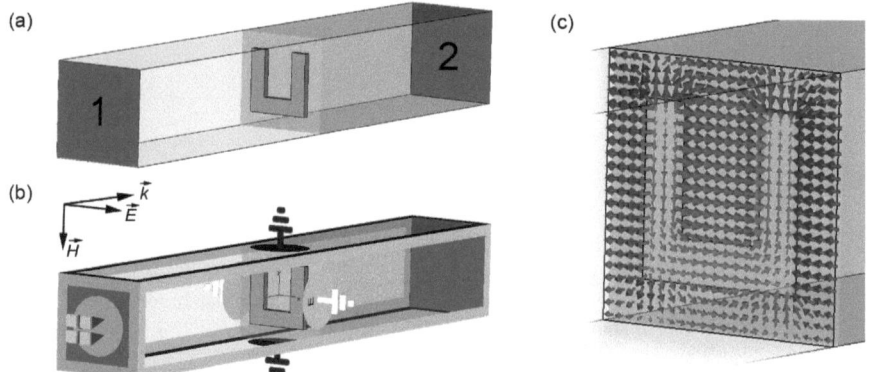

Figure 3.10: Computer-aided design model (CST Microwave Studio) of a split-ring resonator placed on a glass substrate in a vacuum waveguide. **(a)** Geometry which serves for calculation. The port 1, at which a plane wave is launched inside the waveguide, and the opposite port 2, at which the propagated wave is detected, are illustrated. **(b)** To define the polarization of the wave, the vertical component of the electric field (light plane) and the horizontal component of the magnetic field (black plane) are set to zero. The gray front plane indicates open boundary conditions. **(c)** Cut through the SRR plane. The electric fields at the resonance frequency of the split-ring resonator are calculated by using the finite-integration technique.

in a waveguide geometry (shown in Fig. 3.10). The whole waveguide geometry is meshed with a spatial discretization of at least the tenth part of the minimum wavelength to be evaluated in order to provide reliable results. Then, for each mesh cell of the finite-element domain one certain material dispersion is defined. Finally, as proposed in Ref. [139], the discretized version of Maxwell's equations is solved in a leap-frog manner.

The structures are designed by using Microwave Studio's internal computer-aided design interface. For each material, experimental data is approximated by simple models for the frequency range of interest to individually set the permittivity $\varepsilon(\omega)$ of each component. The respective values are listed in Tab. 3.1. Note that dielectrics are considered as non-dispersive materials which is clearly an approximation. Moreover, we slightly modified the Drude parameters for gold and silver as they differ from Refs. [76, 77]. Especially, the collision frequency γ must be increased for thin films due to significant surface roughness. Then, the whole structure is embedded in vacuum and placed on a glass substrate according to real-world experiments.

At one port (front plane in Fig. 3.10(a)) of the waveguide a plane wave is launched into the simulation volume and the output signals are detected at both ports. Because of the finite computation domain, the values of the fields on the boundaries must be defined so that the waveguide appears to be extend infinitely in lateral directions. Additionally, the polarization of the wave must be defined correctly. For this purpose, the tangential components of the $\vec{E}$- and $\vec{H}$-fields are set to zero for the planes perpendicular to the propagation direction in the case of waveguide boundary conditions (see Fig. 3.10(b)). Alternatively, it is also possible to use periodic boundary conditions, where the fields can be detected at certain points in the waveguide volume. However, these conditions restrict the calculations to normal incidence.

Table 3.1.: Material parameters used for numerical calculations. The Drude parameters for silver and gold are taken from Ref. [59]. Note that the latter are chosen differently from those in Tab. 2.2 since the surface roughness of thin films modifies the collision frequency γ.

Element	ω_{pl} (10^{15} Hz)	γ (10^{12} Hz)	ε
Gold (Au)	13.7	85.0	-
Silver (Ag)	13.7	40.7	-
Silica (SiO$_2$)	-	-	2.2
Titania (TiO$_2$)	-	-	4.2
SU-8	-	-	2.4
Air / vacuum	-	-	1.0

Along the propagation direction, open boundary conditions are applied. Note that the distance from the structure to the ports is chosen such that near fields are decayed and do not distort the results.

In addition to the finite-integration technique, we also deployed other simulation software for consistency checks, i.e., 1D and 2D Fourier modal method codes[7] (scattering-matrix approach) based on Refs. [140–143].

3.4.2. Measurement of Optical Spectra

To measure the transmittance and reflectance of the fabricated metamaterials over a wide spectral range, we have access to two Fourier-transform infrared (FTIR) spectrometers (Bruker Equinox 55 and Bruker Tensor 27). Since standard optical components like lenses and beam splitters—which are mostly made of glass—become opaque at around 3 µm wavelength, FTIR spectrometers primarily involve reflective optical components, despite of one beam splitter made of potassium bromide (KBr). Thus, these devices are usable for a very broad spectral range. However, depending on the choice of detector and beam splitter, the FTIR systems Equinox 55 and Tensor 27 provide a limited measuring range from 1 µm to 5 µm and 2 µm to 13 µm, respectively.

The working principle of FTIR spectrometers is based on a Michelson interferometer (see Fig. 3.11) which enables to measure the field autocorrelation[8] $A(t)$. As stated by the Wiener-Khinchin theorem [144], $A(t)$ corresponds to the Fourier transform of the power spectrum $I(\omega)$,

---

[7] The development of the 2D Fourier modal method code and related calculations have been carried out by Sabine Essig at the Institut für Theoretische Festkörperphysik (Karlsruhe Institute of Technology).

[8] The intensity measured by the FTIR detector is given by $I(t) = \int_{-\infty}^{\infty} |E(t') + E(t'-t)|^2 \, dt'$. Expanding $I(t)$ reveals that one of the terms corresponds to the field autocorrelation function $A(t) = \int_{-\infty}^{\infty} E(t')E^*(t'-t) \, dt'$, where $E(t')$ is the electric field and the star denotes the complex conjugate of the time-harmonic function. We see that $A(t)$ is not phase-sensitive. Note that the time response of the detector must be much larger than the duration of the signal $E(t')$ or, alternatively, the recorded signal has to be integrated.

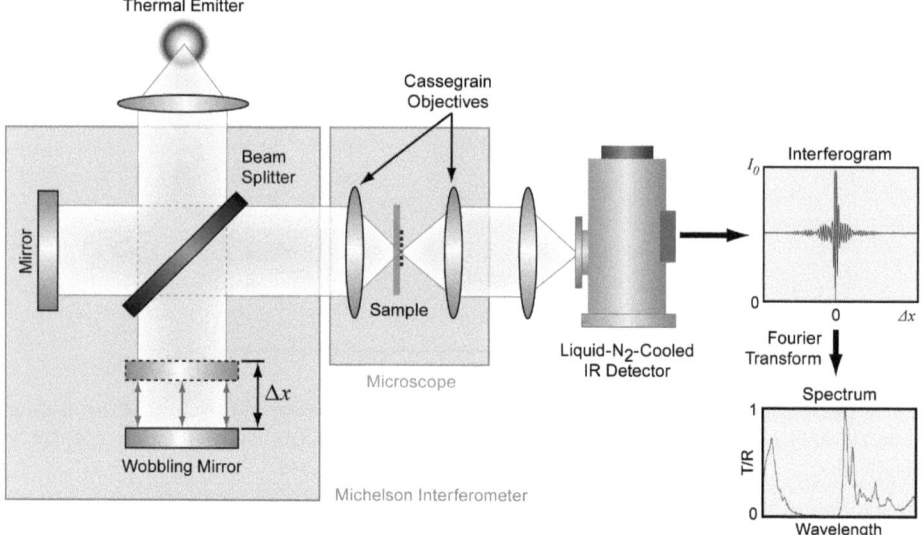

Figure 3.11: Schematic of a FTIR spectrometer assembly. The systems consist of a microscope attached to a Michelson interferometer. Apart from the beam splitter, solely reflective optical components are used since most materials become opaque at frequencies of interest.

i.e.,

$$\mathcal{F}(A(t)) = \mathcal{F}\left(\int_{-\infty}^{\infty} E(t')E^*(t'-t)\,\mathrm{d}t'\right) = |\mathcal{F}(E(t))|^2 = |E(\omega)|^2 \sim I(\omega)\ .$$

This concept becomes clearer as we follow the beam path and analyze the effect of each optical component: Radiation from a thermal emitter (white-light source) is directed to a beam splitter which must be highly transparent in the whole measuring range. Half of the incident light intensity is reflected by a fixed mirror while the rest is reflected by a continuously moving mirror (actually introducing the field autocorrelation). For each wavelength λ, the respective intensity coming out of the interferometer depends on the path difference $2\Delta x$ between both mirror arms, i.e., [84]

$$I_\mathrm{ref}(\Delta x) = 2\left|\frac{S(\lambda)}{4}\right|^2 \left(1 + \cos\left(\frac{2\pi}{\lambda}\Delta x\right)\right)\ ,$$

where $S(\lambda)$ denotes the amplitude spectrum of the electric field coupled into the interferometer. Hence, the intensity of the whole white-light spectrum reads

$$I_\mathrm{ref}(\Delta x) = \int_0^\infty 2\left|\frac{S(\lambda)}{4}\right|^2 \left(1 + \cos\left(\frac{2\pi}{\lambda}\Delta x\right)\right)\,\mathrm{d}\lambda\ .$$

Its Fourier transform can be calculated via

$$\mathcal{F}(I_\mathrm{ref}(\Delta x)) = c\,|S(\lambda)|^2\ ,$$

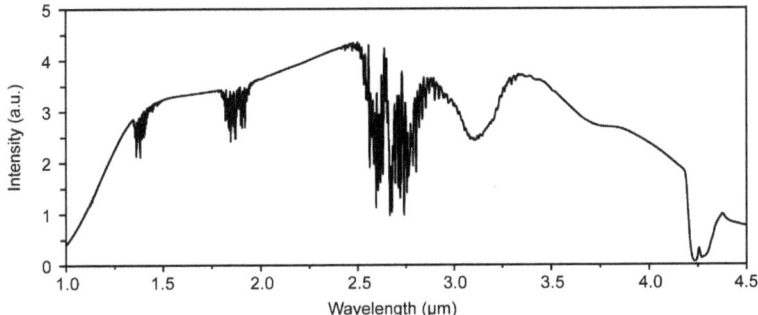

Figure 3.12: Transmittance reference spectrum of the Bruker Equinox 55 Fourier-transform infrared spectrometer scaled in arbitrary units. A 170 μm-thick glass substrate is used as a sample between both Cassegrain objectives (q.v. Fig. 3.11). Absorption bands due to O-H bond (between 2.5 μm and 3.0 μm wavelength), C-H bond (between 3.0 μm and 3.5 μm wavelength), O=C=O bond (at 4.2 μm wavelength), and silica (> 6.0 μm wavelength) resonances are present. Moreover, the sensitivity of the detector (in this case a liquid-nitrogen-cooled indium antimonide (InSb) detector) decreases below 1.5 μm and above 4.0 μm wavelength. However, this loss in signal amplitude can be compensated to a certain extend by averaging of multiple measurements.

with c being a constant.

For the case of detecting the reference spectrum, the calculation stops at this point. If, however, a sample is placed into the beam path, the spectrum of the incident light is modified according to its transmittance $T(\lambda)$, i.e., $I(\Delta x) = T(\lambda) I_{\text{ref}}(\Delta x)$. The respective Fourier transform reads

$$\mathcal{F}(I(\Delta x)) = c T(\lambda) |S(\lambda)|^2 = T(\lambda) \mathcal{F}(I_{\text{ref}}(\Delta x)) \ .$$

The reflectance can be treated in an analog manner. Finally, the spectrum of the metamaterial is given by

$$T(\lambda) = \frac{\mathcal{F}(I)}{\mathcal{F}(I_{\text{ref}})} \ .$$

A single scan of the entire mirror distance takes about one second. To reduce measurement artifacts like thermal fluctuations and vibrations in the laboratory, a helium-neon (HeNe) laser is simultaneously directed through the Michelson interferometer. Its interference pattern is used as a frequency reference to precisely determine the position of the movable mirror.

Fig. 3.12 shows a typical reference spectrum of a near-IR FTIR. It accounts for the emission spectrum of the lamp, the absorption of all media crossed by the light beam, and the sensitivity of the detector. The reference spectrum is measured by focusing the beam onto the surface of the substrate. In particular, we can observe resonant molecular absorption modes of O-H bonds (between 2.5 μm and 3.0 μm wavelength), C-H bonds (between 3.0 μm and 3.5 μm wavelength), and O=C=O bonds (at 4.2 μm wavelength). Moreover, at wavelengths above 6.0 μm (not shown in Fig. 3.12) the 170 μm-thick glass substrate becomes opaque. For these spectral domains, the intensity of the reference beam is relatively low which makes the sampling more sensitive to noise.

As we like to measure spectra of small nanostructures (typically with total lateral dimensions of a few ten microns), we must focus the white-light beam to increase the local intensity. Therefore, both spectrometers are coupled to a purpose-built microscope (Bruker Hyperion 2000) equipped with reflective Cassegrain objectives (magnification: ×36, numerical aperture: 0.5). The regions of interest are selected either by circular apertures or by knife-edges.

Due to the intrinsic geometry of Cassegrain objectives, by default, the samples are illuminated by an oblique incident light beam with a finite opening angle of 15° to 30°. As mentioned, the calculations *via* finite-integration technique can only be performed for normal incidence. This discrepancy can be partly avoided by introducing apertures which mask most of the incident light. By additionally tilting the sample, the beam hits the structure with a minimized opening angle of $< 5°$—which is a fairly good approximation to normal incidence. Alternatively, the Cassegrain objectives can be replaced by calcium fluoride (CaF_2) objectives which show high transmittance up to 8 µm wavelength. However, their magnification is much lower so that only fairly large structures can be characterized.

4. Bi-Anisotropic Three-Dimensional Metamaterials

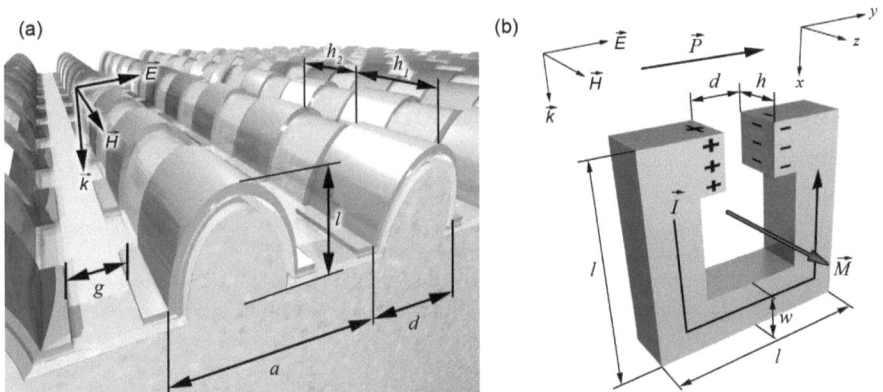

Figure 4.1: **(a)** Artist's view of an array of upside-down 3D split-ring resonators (SRRs) which we intend to examine. Adapted from [109]. **(b)** Illustration of a stylized SRR and parameters used for the analytic calculation. The excitation geometry considered in this section is also depicted. Notably, the incident light is not perpendicular to the SRR's symmetry plane (i.e., the xy-plane). Therefore, the unit cells are not centrosymmetric along the propagation direction which yields a bi-anisotropic optical behavior. Adapted from Ref. [51].

For a first demonstration of the proposed 3D fabrication method presented in section 3.3, i.e., DLW in combination with silver CVD, we fabricated an array of separated 3D SRRs like shown in Fig. 4.1(a) [32, 109, 112, 113]. Here, the excitation geometry is fundamentally different to *planar* SRR arrays [7, 24–27] for which we already derived analytic expressions in section 2.4. It rather relates to the case depicted in Fig. 2.4(c), where the structure couples to both the electric and the magnetic component of light. To provide an insight into the underlying physics, we proceed on the assumption that the coupling between adjacent SRRs is negligible. Hence, the derived optical properties of an individual SRR can be directly related to the properties of a metamaterial consisting of an array of these building blocks. Beyond that, we simplify the unit cell by rectifying the outer shape and consider the SRRs to be excited from the bottom. This leads to a more instructive representation shown in Fig. 4.1(b). By regarding this illustration, we realize that the electric field vector $\vec{E}$ of the incident light induces a polarization $\vec{P}$. The charge separation at the slit causes a flowing current $\vec{I}$ which, in turn, induces a magnetization $\vec{M}$. Due to the symmetry of the SRR, the induced magnetic dipole moment is oriented perpendicular to the exciting electric field and parallel to the incident magnetic component $\vec{H}$. In the same way, we could also start from the incident magnetic component of light and will finally end up with an induced polarization which is perpendicular to the excitation. Since the induced field components include an angle $\varphi = 90°$ with the fields which excited them, the underlying configuration is concerned with "pure" bi-anisotropy previously discussed in section 2.8.1. This is additionally confirmed by the unit cell—and hence the respective metamaterial—not being centrosymmetric along the propagation direction of light.

In the following section, we will re-derive an analytic SRR model in quasi-static approximation accounting for bi-anisotropy [51]. The resulting expressions will also be checked by numerical calculations to provide confidence for following discussions. This initial theoretical approach will help to understand the measured spectra in section 4.2 as well as the calculated dispersion of the material parameters.

4.1. Analytic Model of a Bi-Anisotropic Split-Ring-Resonator Array

For the sake of deriving explicit expressions for $\vec{P}$ and $\vec{M}$, we start with Kirchhoff's law like previously in section 2.4. However, this time the excitation geometry (see Fig. 4.1) is different. To keep track of the full problem, we split our discussions into the contributions of Faraday's induction law and the voltage drop over the slit capacitor. Furthermore, we consider only the field vector components of interest.

Contribution of the Faraday induction law: The magnetization induced by the magnetic field component can be directly obtained by using (2.34), i.e.,

$$M_z = \chi_{m,zz}(\omega) H_z = (\mu_{zz}(\omega) - 1) H_z$$
$$= \left(\frac{f\omega^2}{\omega_0^2 - \omega^2 - 2i\gamma\omega} \right) H_z \;. \tag{4.1}$$

In (4.1), we introduced the damping $\gamma = R/(2L)$, the LC eigenfrequency ω_0, the SRR volume filling fraction $f = l^2 h/V$, and the unit volume $V = \prod_{i=1}^{3}(a_i)$. Due to bi-anisotropy the Faraday law also results in a time-harmonic polarization

$$P_y = \frac{d}{V} \int I \, \mathrm{d}t$$

which is related to the magnetic field *via* the cross-term parameter

$$\begin{aligned}
\xi_{yz}(\omega) &= c_0 \frac{P_y}{H_z} \\
&\stackrel{(2.33)}{=} c_0 \frac{\mathrm{d}I}{(-i\omega)V} \frac{\mu_0 l^2 \omega^2}{(\omega_0^2 - \omega^2 - 2i\gamma\omega) IL} \\
&\stackrel{(2.28)}{=} c_0 \frac{hl^2 d}{l^2 V} \left(\frac{i\omega}{\omega_0^2 - \omega^2 - 2i\gamma\omega} \right) \\
&\stackrel{(2.35)}{=} c_0 \frac{d}{l^2} \left(\frac{if\omega}{\omega_0^2 - \omega^2 - 2i\gamma\omega} \right) \;.
\end{aligned} \tag{4.2}$$

The polarization arising from the magnetic component of the incident light is given by

$$P_y = \frac{1}{c_0}\xi_{yz}(\omega)H_z = \frac{d}{l^2}\left(\frac{if\omega}{\omega_0^2-\omega^2-2i\gamma\omega}\right)H_z \ . \tag{4.3}$$

Notably, P_y is phase delayed by 90° with respect to the exciting magnetic field H_z. Besides that, the polarization reveals a similar resonance behavior around the LCR eigenfrequency as the magnetization. As intuitively expected, the absolute value of the polarization changes proportional to the slit width d of the SRR.

Contribution of the voltage drop: The voltage drop over the slit capacitor arises from the incident $\vec{E}$-field and is given by $U_{\text{ind}}=E_y(t)d$. Assuming again time-harmonic fields results in

$$U_{\text{ind}} = -\frac{\partial \Phi_B}{\partial t} = -\mu_0 l^2(-i\omega)H_z \ . \tag{4.4}$$

Hence, the derivation of M_z caused by the voltage drop reads

$$\zeta_{zy}(\omega) = \mu_0 c_0 \frac{M_z}{E_y}$$

$$= \mu_0 c_0 \frac{Il^2}{V} \frac{d}{U_{\text{ind}}}$$

$$\stackrel{(4.4)}{=} \mu_0 c_0 \frac{Il^2}{V} \frac{d\,\mu_0 l^2}{(i\omega)\mu_0 l^2} \frac{\omega^2}{(\omega_0^2-\omega^2-2i\gamma\omega)IL}$$

$$\stackrel{(2.28)}{=} c_0 \frac{d}{l^2}\left(\frac{-if\omega}{\omega_0^2-\omega^2-2i\gamma\omega}\right)$$

$$\stackrel{(4.2)}{=} -\xi_{yz}$$

$$\Rightarrow M_z = \frac{1}{\mu_0 c_0}\zeta_{zy}(\omega)E_y$$

$$= \frac{d}{\mu_0 l^2}\left(\frac{-if\omega}{\omega_0^2-\omega^2-2i\gamma\omega}\right)E_y \ . \tag{4.5}$$

For the polarization caused by the voltage drop, we finally obtain

$$\varepsilon_{yy} = 1+\frac{P_y}{E_y}$$

$$= 1+\frac{dI}{(-i\omega)V}\frac{d}{U_{\text{ind}}}$$

$$\stackrel{(4.4)}{=} 1+\frac{d^2 I}{(-i\omega)V}\frac{\mu_0 l^2}{(i\omega)\mu_0 l^2}\frac{\omega^2}{(\omega_0^2-\omega^2-2i\gamma\omega)IL}$$

$$\stackrel{(2.28)}{=} 1+\frac{d^2}{\mu_0 l^4}\left(\frac{f}{\omega_0^2-\omega^2-2i\gamma\omega}\right)$$

$$\Rightarrow P_y = \varepsilon_0 \chi_{e,yy}(\omega)E_y$$

$$= \frac{\varepsilon_0 d^2}{\mu_0 l^4}\left(\frac{f}{\omega_0^2-\omega^2-2i\gamma\omega}\right)E_y \ . \tag{4.6}$$

In contrast to the contribution of the Faraday induction law, the 90° phase delay now occurs for the magnetization M_z. Apparently, a composite medium consisting of the discussed SRR array is reciprocal because of $\xi_{yz} = -\zeta_{zy}$. If the SRR shown in Fig. 4.1 is excited from the opposite side, i.e., the wave vector is anti-parallel, the induced current $\bar{I}$ is also inverted. Hence, $+\xi_{yz} \rightarrow -\xi_{yz}$, whereas ε_{yy} and μ_{zz} do not change their sign.

The derived material parameters *versus* the normalized frequency ω/ω_0 are shown in Fig. 4.2. Although both $\text{Re}(\varepsilon) < 0$ and $\text{Re}(\mu) < 0$ are simultaneously negative in the spectral region highlighted by the gray bar, this does not essentially lead to $\text{Re}(n) < 0$, not even if damping is neglected (i.e., $\gamma/\omega_0 = 0$). Referring to the discussions of section 2.8.1, especially Fig. 2.12(b), this is a distinct property of bi-anisotropic media. Another typical signature is found for the spectrum in Fig. 4.2(e). If light is impinging from the top (i.e., propagating along the $+x$-direction), we obtain a reflectance denoted by R_+, whereas the reflectance for light impinging from the bottom (i.e., propagating along the $-x$-direction) is given by R_-. Notably, the reflectance depends on direction which can be associated with the bi-anisotropic impedance given by (2.52). The transmittance T, however, is equal in both directions ($T = T_- = T_+$) demonstrating that the SRR array is indeed reciprocal.

We compared the analytic model with time-domain calculations performed by Microwave Studio. Referring to Fig. 4.1(a), we used the following geometrical parameters: Periodicity $a = 1000$ nm, SRR slit width $d = 570$ nm, height $l = 525$ nm, SRR thickness $h_1 = 550$ nm, spacer thickness $h_2 = 450$ nm, groove width $g = 270$ nm. Notably, these are the feature sizes obtained from evaluation of electron micrographs of the fabricated structures to be discussed in the following section. The qualitative agreement between the numerical calculations and the analytic results from Fig. 4.2 is very good. Especially, the direction dependence of the reflectance $R_\pm$ is nicely reproduced by numerics. Note that the calculated SRRs are upside-down—compared to the geometry assumed for the analytic model. Therefore, they are excited from the bottom leading to a mirroring of the cross-term's curve (at the $\xi = 0$ axis) in Fig. 4.3(c) compared to Fig. 4.2(c). In a nutshell, the analytic quasi-static model seems to comprise the most important optical characteristics as similar signatures can be found in both cases.

Let us, for a moment, suppose that a second slit is introduced to the SRR, opposite to the first one. At first view, it becomes evident that inversion symmetry is recovered. While neglecting retardation effects, the voltage drop over the second slit is opposite in sign to the first one. Thus, neither magnetization nor polarization is induced by the electric field, whereas a magnetic field component normal to the SRR plane can still induce a circulating and oscillating current, leading to a magnetic dipole moment. However, the electric dipole moment of the second slit is opposite to that of the first slit. Hence, no electric polarization results from the magnetic field of the incident light. It directly follows from our model that under these conditions the cross-term parameter ξ_{yz} is strictly zero and $\varepsilon_{yy} = 1$ while $\mu_{zz} \neq 1$.

4.2. Fabrication and Optical Characterization

So far, our investigations have been restricted to theoretical analyses to gain a deeper insight to the fundamental behavior of a bi-anisotropic array of 3D SRRs. The next step is to fabricate the corresponding samples and to see whether the experimental characterization coincides with the results from the prior section. Because of the utilized fabrication method, we additionally obtain descendants of the intended 3D SRR structure whose unit cells are electrically connected along

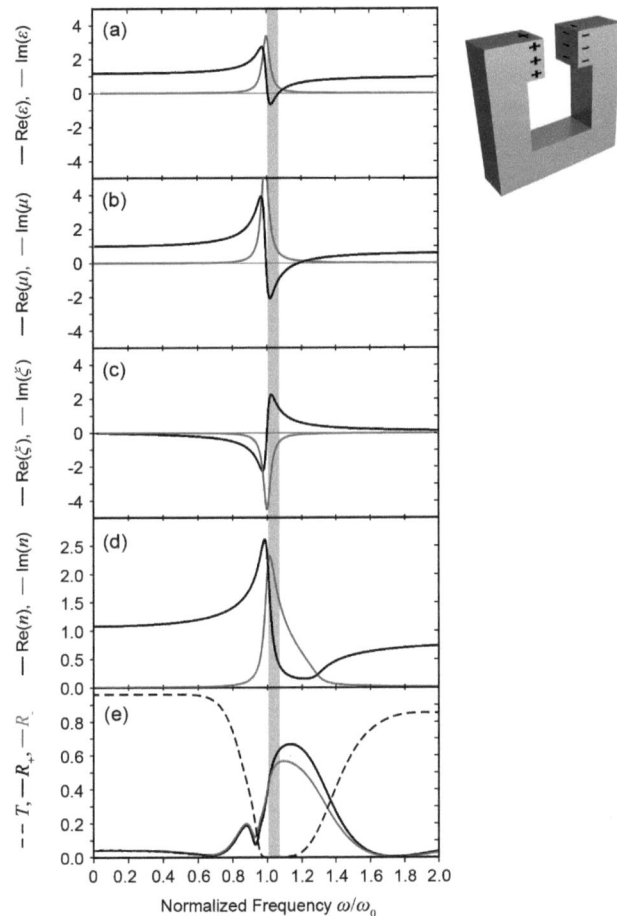

Figure 4.2: Exemplary results of the derived analytic split-ring resonator model (for geometry see Fig. 4.1) for the parameters $\gamma/\omega_0 = 0.05$, $f = 0.3$, and $dc_0/l^2 = 0.75\,\omega_0$. (a) Electric permittivity ε, (b) magnetic permeability μ, (c) cross-term parameter ξ, and (d) refractive index n *versus* the normalized frequency ω/ω_0. Real (imaginary) parts of these complex quantities are shown in black (gray). (e) Normal-incidence transmittance $T = T_+ = T_-$ and reflectances R_+ (light impinging from the top, i.e, the $+x$-direction) and R_- (light impinging from the $-x$-direction) for a slab of material with parameters as in (a)–(d). The slab thickness is taken as $c_0\pi/\omega_0$. Taken from Ref. [51].

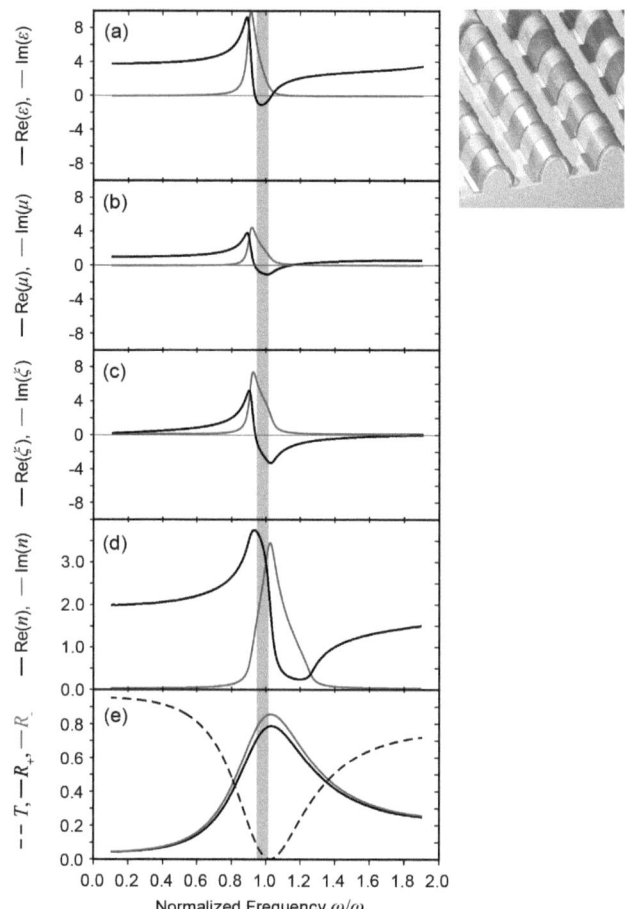

Figure 4.3: Calculated spectra and retrieved material parameters *versus* the normalized frequency ω/ω_0 of an array of upside-down split-ring resonators shown in the inset. For normalization, the LC eigenfrequency was estimated from the spectral position of the resonance (i.e., the dip in transmittance) to be $\omega_0 = 70$ THz. The parameters of the structure are listed in the text. **(a)** Electric permittivity ε, **(b)** magnetic permeability μ, **(c)** cross-term parameter ξ, and **(d)** refractive index n retrieved from the equations presented in section 2.8.1. Real (imaginary) parts of these complex quantities are shown in black (gray). **(e)** Normal-incidence transmittance $T = T_+ = T_-$ and reflectances R_+ (light impinging from the top) and R_- (light impinging from the bottom). The regions $0 < \omega/\omega_0 < 0.1$ and $1.9 < \omega/\omega_0 < 2.0$ are intentionally masked due to artifacts of the time-domain solver at frequencies near zero and close to Wood anomalies, respectively.

Figure 4.4: Oblique-view electron micrographs of the fabricated (a) 3D split-ring resonator (SRR) array and (b)–(d) corresponding variations. The insets illustrate the intended designs. The corrugated-metal surface in (d) is realized by direct laser writing and subsequent silver metallization *via* chemical vapor deposition. The structures in (a)–(c) require additional post-processing by means of focused-ion beam milling. The geometrical parameters evaluated from these micrographs are as follows: Periodicity $a = 1000$ nm, SRR slit width $d = 570$ nm, height $l = 525$ nm, SRR thickness $h_1 = 550$ nm, spacer thickness $h_2 = 450$ nm, groove width $g = 270$ nm. The definition of these geometrical parameters can be found in Fig. 4.1(a). Adapted from Ref. [109].

the two in-plane directions (*q.v.* Fig. 4.4(b)–(d)) [109]. To distinguish the fabricated structures, they are named after their sub-figure indication in Fig. 4.4, i.e., "case (a)" relates to the 3D SRR design shown in Fig. 4.4(a), "case (b)" to Fig. 4.4(b), and so forth. This nomenclature will be used throughout the rest of this chapter.

Fabrication of the 3D SRR array starts with a glass substrate covered with a 2 μm-thick fully polymerized SU-8 film. This preliminary layer is used to prevent tearing of the silver layer due to thermal stresses between the glass substrate and the SU-8 structure at high temperatures. Subsequently, another SU-8 film is spun-on, structured *via* DLW, post-baked, and developed. By using the DLW's tilt-correction feature, substantial lateral gradients in height within the sample footprint (here 50 μm ×50 μm) can be avoided. Afterwards, the resulting polymer template is protected by adding a thin titania layer *via* ALD and is finally coated with a 34 nm-thick silver layer using CVD. This procedure results in the corrugated metal surface depicted in Fig. 4.4(d). The other structures shown in Fig. 4.4(a)–(c) have been post-processed by FIB milling. We use a FIB / scanning-electron microscopy system (Zeiss 1540 XB) operating with gallium ions (Ga$^+$) at 30 keV. Milling along the grooves has been performed manually by imaging the trenches with an ion beam of 20 pA. For orthogonal cuts in Fig. 4.4(a) and (b), we have utilized an ELPHY Plus nanolithography system (by Raith GmbH) for automatically controlling the FIB milling with a beam current of 100 pA. Resulting feature sizes can be measured from electron micrographs and are listed in the caption of Fig. 4.4. Notably, the fabricated structures come fairly close to the desired ideal (see insets).

The normal-incidence transmittance spectra (left column of Fig. 4.5) have been measured using the Bruker Equinox 55 FTIR. All results are for linear polarization of the incident light oriented perpendicular to the grooves. The spectra are normalized to the transmittance of a bare glass substrate. In Fig. 4.5, the column on the right-hand side shows the related calculations for direct comparison with the theoretical ideal. For this purpose, the Maxwell equations have been solved numerically for one 3D unit cell with waveguide boundary conditions using the time-domain solver of Microwave Studio. Furthermore, we checked the results for the symmetric cases, i.e., cases (c)–(d), with a self-made Fourier modal method code by employing an expansion of the fields in 801 plane waves. The 2D-symmetric unit cell has been subdivided into 524 layers along the propagation direction of the incident light wave and calculated by using periodic boundary conditions. For all calculations, we used the silver parameters from Tab. 3.1. The refractive indices of SU-8, glass, and titania have been taken as denoted in section 3.3.2.

The experimental results are well reproduced by our calculations. Especially, the transmittance minima which correspond to the excitation of a magnetic-dipole resonance (compare gray shadings of the left and the right column) agree fairly well. Remaining quantitative deviations are likely due to simplifications in the considered model geometry and / or due to fabrication imperfections. In particular, it is known that FIB milling tends to introduce gallium contamination which deteriorates the optical properties.

The transmittance spectra in Fig. 4.5 clearly reveal that the resonance positions of the initially investigated 3D SRR array (case (a)) and descendants differ considerably. To disclose the origin of the observed characteristics, we should survey the differences between the fabricated structures in a stepwise manner.

4.3. From Isolated Split-Ring Resonators to Corrugated Metal Surfaces

Electrically connecting the isolated 3D SRRs of case (a) perpendicular to the grooves leads to the design of case (b). This transition looks like a minor modification but has far-reaching consequences for the optical response. At first view, a strong blue-shift ($\Delta\lambda \approx 1.8\,\mu$m) of the magnetic resonance compared to case (a) is observed from the transmittance spectra in Fig. 4.5(b) and (f), respectively. To identify the origin of this variation, we retrieve the bi-anisotropic optical parameters. The respective results in Fig. 4.6 show that the dispersion of the permittivity $\varepsilon(\omega)$ has now a Drude-like progression which is overlapped by a resonance at approximately 100 THz. This fact can be explained by realizing that the electrical connection has formed a diluted metal whose plasma frequency is in the range of 60 THz. Note that the frequency axis in Fig. 4.6 is not normalized and the scales of the material parameters are different to those in Fig. 4.3. Remarkably, the cross-term parameter's progression $\xi(\omega)$ is similar

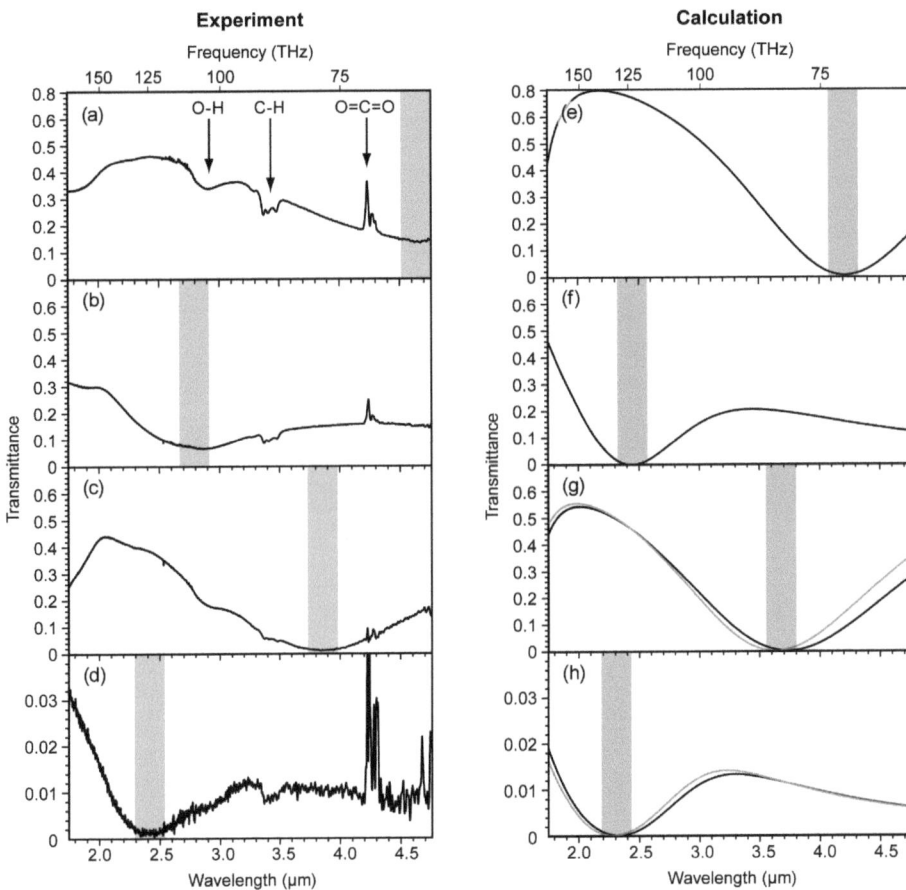

Figure 4.5: **Left column:** Measured normal-incidence transmittance spectra for the four samples shown in Fig. 4.4 (ordered consistently). The gray shadings highlight the magnetic resonance for each case. Due to low transmittance, the vertical scale in (**d**) is chosen differently. Artifacts resulting from chemical absorption lines are indicated in (**a**). **Right column:** Calculated spectra referring to the measuring results in the left column. In (**g**) and (**h**) finite-integration technique calculations (black) are compared with Fourier modal method calculations (gray) using the same geometrical parameters. Note that the scales in both columns are identical, thus allowing for a direct comparison. Adapted from Ref. [109].

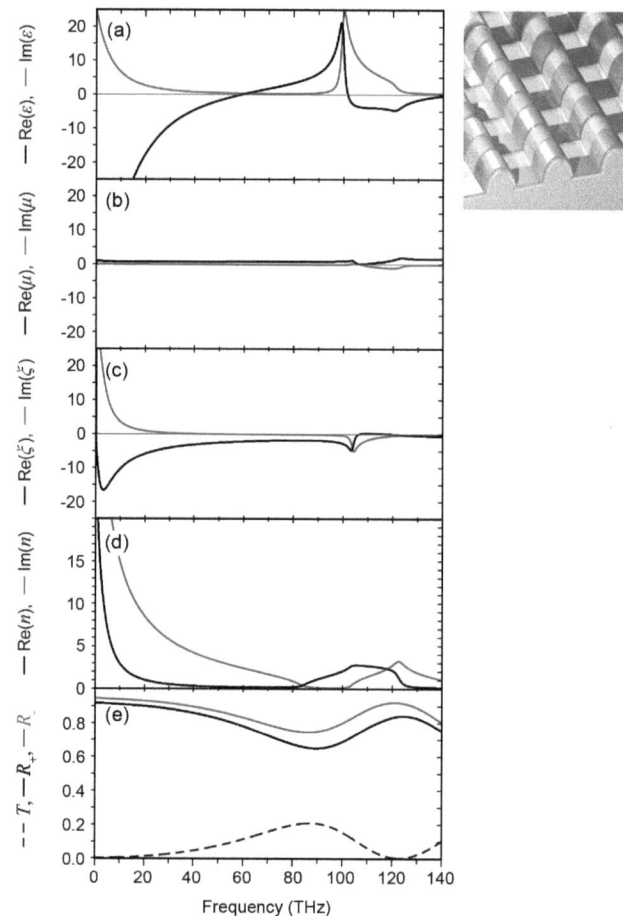

Figure 4.6: Calculated spectra and retrieved material parameters *versus* frequency of case (b), i.e., the structure design shown in Fig. 4.4(b). The parameters of the structure have been taken as listed in the caption of Fig. 4.4. (a) Electric permittivity ε, (b) magnetic permeability μ, (c) cross-term parameter ξ, and (d) refractive index n retrieved from the equations presented in section 2.8.1. Real (imaginary) parts of these complex quantities are shown in black (gray). (e) Normal-incidence transmittance $T = T_+ = T_-$ and reflectances R_+ (light impinging from the top) and R_- (light impinging from the bottom). Note that the transmittance spectrum is identical to Fig. 4.5(f), but scaled differently.

to the analytic model of the upright SRR array which we discussed in section 4.1. This suggests that the functional unit cell has rotated by 180°. To verify this assumption, we have plotted the calculated current density of a 2D cross section for the cases (a) and (b) (see Fig. 4.7). Due to symmetry, the current densities of the cases (c) and (d) look rather similar.

For the case (a) of separated SRRs, the fundamental LCR resonance is clearly that of an upside-down "U" (see Fig. 4.7(a)). The oscillating circulating currents and, therefore, the magnetic dipoles are induced by the voltage drop over the two ends of the SRR's capacitor part and, simultaneously, by the magnetic field normal to the drawing plane. Case (b) is a bit more sophisticated. Here, the right-hand side plate of one SRR is electrically connected to the left-hand side plate of the SRR to its right. As the structure is excited from the top *via* a plane electromagnetic wave under normal incidence, all unit cells are forced to oscillate in phase. This means that they all have the same potential difference between their left and right-hand side plate. Combined with the discussed electrical connection of neighboring SRRs, this is essentially equivalent to short-circuiting each plate capacitor, i.e., each SRR becomes a closed ring. Successively closing the gap of a "U" towards an "O" corresponds to a diverging capacitance and, thus, to a fundamental LCR eigenfrequency approaching zero. In this quasi-static case shown in Fig. 4.7(b) (at frequencies of around $1\,\text{THz} \ll 122.5\,\text{THz}$), strong circulating currents lead to strong magnetic dipoles excited by the incident light field. This is contrary to what is known for the separated SRRs. The situation can also be interpreted as two poles of a battery which are applied at the left and right-hand side end of the structure. Clearly, the voltage drop between the two ends leads to an electric current flowing from the left to the right. Interestingly, the current will create local magnetic dipoles in the almost closed metallic loops. As the structure is asymmetric along the propagation direction, the neighboring magnetic-dipole moments do not cancel and a net magnetization of the metamaterial structure remains even in the static limit. For electrically separated SRRs, this continuous current flow is obviously not possible. Thus, the resonance in transmittance observed in Fig. 4.5(b) and (f) is not the fundamental magnetic resonance but rather a higher-order mode. The fundamental mode of the system shifts towards zero frequency (infinite wavelength) when case (a) passes into case (b).

The transition between case (c) and (d) can be treated equally. Here, compared to cases (a) and (b), the SRR thickness is merely enlarged along the groove direction to form continuous wires. From this, we expect a higher resonance frequency due to a decreased kinetic inductance (see section 2.4). In fact, this tendency is clearly observed in the measured optical spectra (*q.v.* Fig. 4.5(a) and (c) as well as Fig. 4.5(b) and (d)). Notably, the effective medium limit is not violated since the induced currents do not have a vector component parallel to the grooves. Hence, the geometry can be regarded as an array of SRRs placed tightly next to each other leaving no gaps.

4.4. Negative-Index Bi-Anisotropic Metamaterial

In contrast to isotropic structures, the realization of a negative refractive index can be suppressed for bi-anisotropic designs by additional cross-term parameters, like shown in Fig. 2.12. This actually occurred for the 3D SRR array and its descendants. However, we also know from theory that bi-anisotropy does not inherently forbid $\text{Re}(n) < 0$. Therefore, we have tried to

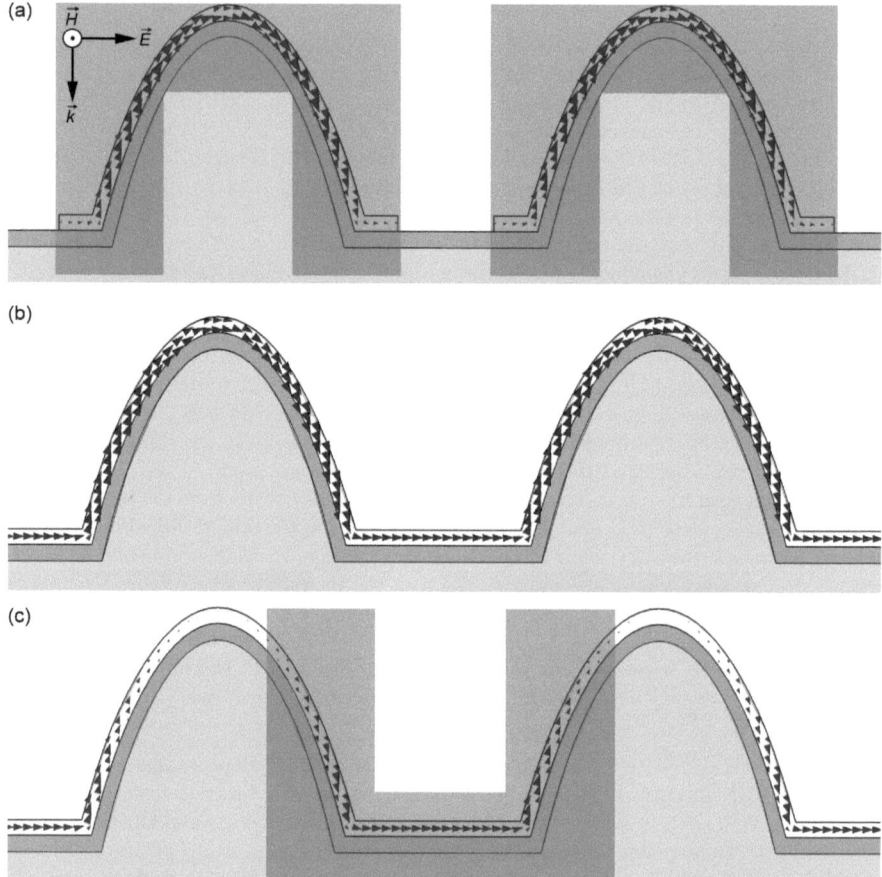

Figure 4.7: Snapshots of the ohmic current density (arrows) inside the metal layer calculated with finite-integration technique. For clarity, a second unit cell is added to the calculated volume. (a) Current distribution of case (a) at the resonance frequency of 70.0 THz (4.25 µm wavelength). (b) Current distribution of case (b) near the fundamental resonance frequency at 1.0 THz (300 µm wavelength), i.e., essentially the static limit. (c) Current density for the higher-order resonance of case (b) at the resonance frequency of 122.5 THz (2.45 µm wavelength). The shaded U-shaped areas in (a) and (c) illustrate our intuitive interpretation of the unit cells' orientation. Reproduced with permission from Ref. [109].

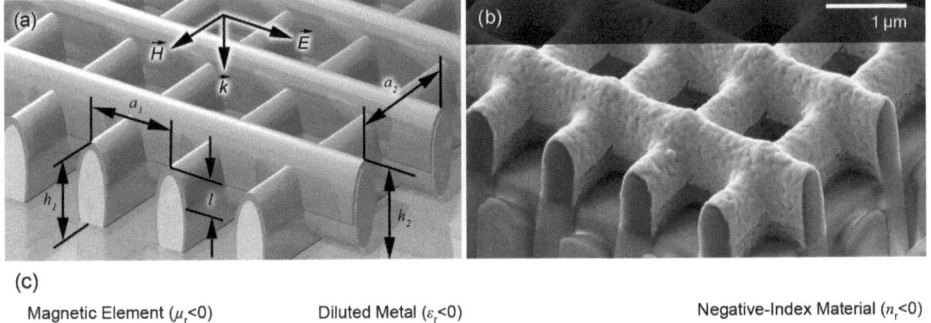

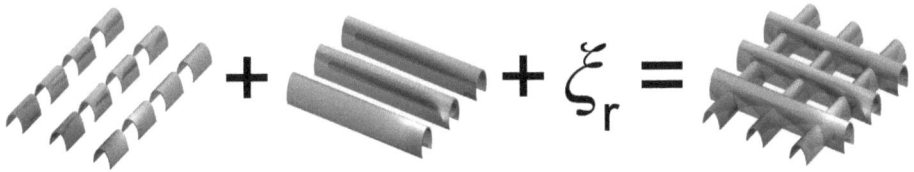

Figure 4.8: Metamaterial design with a negative refractive index. (**a**) Artist's view of the structure design. The white regions are related to the polymer (SU-8) located on a glass substrate. The sidewalls of the polymer are first encapsulated by silica *via* pulsed layer deposition and subsequently coated with silver. The polarization of the incident electromagnetic field and the definition of the geometrical parameters (values are given in the text) are illustrated. (**b**) Oblique-view electron micrograph of the structure fabricated by using direct laser writing and silver shadow evaporation. The structure has been cut *via* a focused-ion beam (FIB) to reveal its interior. The complicated features visible underneath the glass-substrate surface are due to the FIB cutting and, hence, not relevant. (**c**) According to Fig. 2.11, the negative-index structure is composed of a magnetic element, represented by upside-down split-ring resonators. The elevated wires act as a diluted metal. Additionally, a cross-term parameter must be considered which also influences the refractive index for bi-anisotropic materials. (a) and (b) adapted from Ref. [110].

enforce this property by starting from case (a) and introducing additional wires perpendicular to the grooves. The idea behind this approach is based on the possibility to tune the diluted metal density by varying the periodicity of the elevated wires. This allows us to influence the spectral position of the resonance of $\varepsilon(\omega)$.

The corresponding sample design and the realized structure are illustrated in Fig. 4.8(a) and (b), respectively [110, 111, 113]. First, we fabricated a polymer (SU-8) template by using DLW. Next, this template has been coated with a thin layer of silica using a PLD process and metallized *via* electron-beam shadow evaporation of silver under high vacuum. The surface normal and the axis of evaporation included a fixed angle of 65°. The utilized metallization process is highly *anisotropic*. Thus, bringing the azimuth angle to four different positions during evaporation results in coated SU-8 sidewalls but, e.g., the glass substrate remains uncoated. Note that we did not mention electron-beam shadow evaporation in chapter 3 since it cannot be used in combination with DLW to realize bulk metamaterials. It was rather used for practical purposes

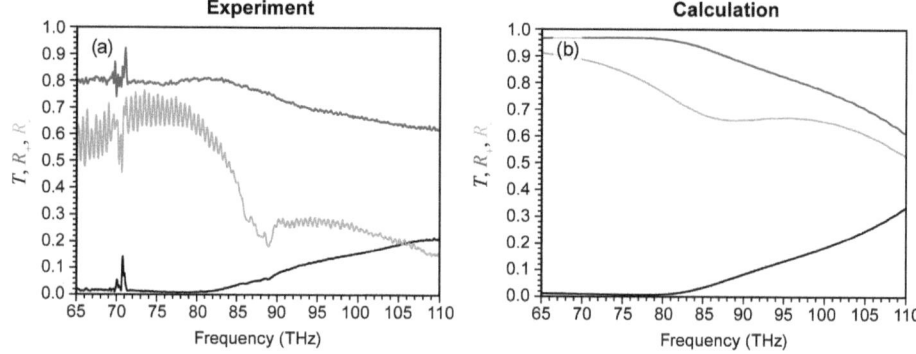

Figure 4.9: Measured linear-optical normal-incidence transmittance (black) and reflectance spectra of the structure shown in Fig. 4.8. Note that the reflectance taken from the air side R_+ (gray) and taken from the glass-substrate side R_- (light gray) are distinctly different. Adapted from Ref. [110].

to circumvent the metallization of the substrate bottom which is an inherent constraint of isotropic coating techniques like CVD.

The upside down "U" parts can be viewed as SRRs which deliver the magnetic-dipole response. The intentionally elevated and elongated metal parts parallel to the incident electric field vector deliver the negative permittivity (see Fig. 4.8(c)). For our analysis, we used a structure with a SRR periodicity of $a_1 = 1.20\,\mu$m, an elevated wire periodicity of $a_2 = 1.70\,\mu$m, rod heights of $h_1 = 1.05\,\mu$m, $h_2 = 1.25\,\mu$m, and a SRR height of $l = 0.88\,\mu$m. The silica film thickness is 35 nm and the silver film thickness is 32 nm on all rod sides and 51 nm on top of the rods.

Measured transmittance and reflectance spectra are shown in Fig. 4.9(a). The transmittance spectrum exhibits a minimum at around 78 THz (i.e., 3.85 μm wavelength). The peaks at around 70 THz are due to carbon dioxide absorption lines in the spectrometer and, hence, an artifact of the measurement. Notably, the normal-incidence reflectance spectrum taken from the air side (gray) is substantially different from that taken from the glass-substrate side (light gray), whereas the transmittance (black) is the same for both sides within experimental uncertainty. This aspect is an immediate consequence of the overall structure being bi-anisotropic. The rapid oscillations are due to Fabry–Pérot interferences in the 170 μm-thick glass substrate.

Next, we compared our experimental results with theory to further elucidate the physics underlying the measured optical spectra. Calculated transmittance and reflectance spectra, directly comparable with experiment, are depicted in Fig. 4.9(b). We used again the finite-integration technique and the material parameters denoted in Tab. 3.1. The qualitative trend of the spectrum agrees well with experiment. In particular, theory also shows a minimum of transmittance at 78 THz. Remaining discrepancies between experiment and theory are very likely due to slight imperfections in sample fabrication (see Fig. 4.8(b)). Importantly, theory reproduces that the normal-incidence reflectance spectra taken from the air and the glass-substrate side are different.

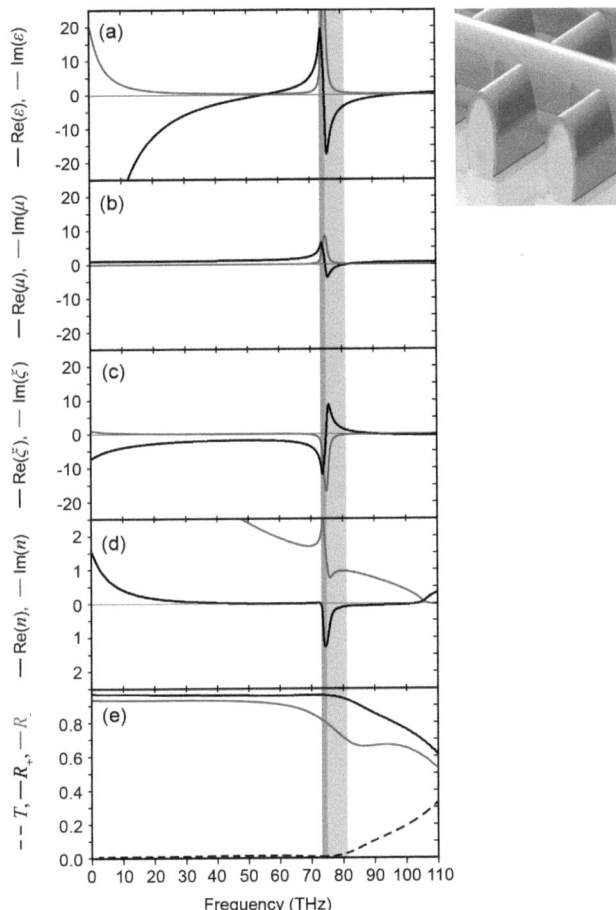

Figure 4.10: Calculated optical response of the bi-anisotropic negative-index metamaterial for normal incidence of light. (a) Retrieved electric permittivity ε, (b) magnetic permeability μ, (c) bi-anisotropy parameter ξ, and (d) refractive index n. The corresponding real parts are shown in black, the imaginary parts in gray. The shaded backgrounds aim at clarifying the origin of the negative real part of n. (e) Transmittance and reflectance spectra for both propagation directions. Adapted from Ref. [110].

The effective metamaterial parameters have been retrieved for a bi-anisotropic slab of thickness $d=1.25\,\mu\mathrm{m}$ on a glass substrate (see Fig. 4.10). In the considered spectral range, a description in terms of an effective medium is justified because the lattice constant is smaller than half the resonance wavelength. Indeed, a negative refractive index is observed from around 77 THz to 83 THz. The negative refractive index is partly connected to $\mathrm{Re}(\varepsilon), \mathrm{Re}(\mu) < 0$ (light shaded area) and partly to $\mathrm{Re}(\xi) < 0$ (dark shaded area). The related imaginary part of n can be translated into a maximum figure of merit ($q.v.$ (2.37)) of FOM = 1.3, which is comparable to double fishnet-type negative-index photonic metamaterials made via electron-beam lithography. This FOM value also clarifies that the very low transmittance in the negative-index region is related to a large impedance mismatch with respect to air rather than being caused by losses.

4.5. Interim Result

In summary, we have investigated an array of upside-down SRRs (and modifications) as an example for a 3D metamaterial fabricated by DLW and silver CVD. The physics of the structure can be understood by an analytic model which accounts for bi-anisotropy. Importantly, the numerical calculations can be accommodated to the characteristics of the fabricated structures. Hence, we conclude that our proposed fabrication method is appropriate to create 3D metamaterials of high optical quality.

Although bi-anisotropy possibly suppresses a negative refractive index, a structure was realized whose cross-term parameter ξ even helps to broaden the spectral range of $\mathrm{Re}(n) < 0$. However, it is evident from Fig. 4.10 that this effect accompanies large values of $\mathrm{Im}(n)$ in the dark gray region. Consequently, the fabricated metamaterial shows an unintentionally high damping of the incident light. To circumvent this problem, we should favor unit cells which are isotropic or anisotropic so that $\xi = 0$ is recovered. Nevertheless, we accomplished a Figure of Merit greater than one which is comparable to recently reported planar metamaterials.

Remarkably, all fabricated 3D metamaterials presented in this section cannot be considered as bulk since the incident light only interacts with a single functional layer. Thus, the optical features are mainly dominated by surface effects. This directly motivates the discussion of the next chapter, where we will present first promising attempts to realize bulk photonic metamaterials.

5. Towards Bulk Photonic Metamaterials

The fabrication techniques, introduced in chapter 3, comprise some intrinsic constraints which must be taken into account in the design process of bulk metamaterials. Hence, in the majority of cases, it is not possible to adapt theoretical drafts [146–151] suggested for metamaterials in the microwave regime, despite their promising properties. Problems concerning the fabrication are encountered in the following situations:

(i) The structure involves locally separated metallic elements along stabilizing connections [146, 147, 149]. Such features are often used to provide a capacitive part required to mimic an LCR circuit. Referring to the fabrication *via* DLW and CVD, this constraint can only be eliminated by establishing local functionalization. Importantly, the functional layer must survive the environmental conditions during the deposition process, i.e., it must be chemically stable at high temperatures and low pressures. Furthermore, the structure-supporting glass substrate is also metallized when using isotropic coating processes like CVD, electroless plating or ALD. If the surface metallization is unintended, it could be removed by a lift-off process. However, related efforts have not been made, yet.

Referring to the fabrication *via* DLW and electroplating, we might overcome this problem by alternating deposition of noble metals (e.g. gold) and easily oxidizable metals (e.g. aluminum).

(ii) The smallest and largest features of the metamaterial unit cell differ greatly in size [150, 151]. Since DLW limits the size of the smallest functional elements to some hundreds of nanometers, the total size of the unit cell must be increased to a certain extent. This often shifts the magnetic resonance to much lower frequencies and / or decreases the λ/a ratio. In the latter case, we run the risk of exceeding the effective medium limit.

Before we published our 3D fabrication method [32, 33], the listed constraints were unknown to the metamaterial community. Hence, it took some time until first compatible proposals like the corrugated-wire structure [145] (see Fig. 5.1(a)) came up. The latter consists of four gold meandering wires successively rotated by 90° to form a unit cell as shown in Fig. 5.1(b). For clarity, the unit cell is depicted from different points of view. Non-isotropic planar versions of this design have been investigated recently [152, 153].

Microwave Studio calculations and an optical parameter retrieval [145] have revealed that the corrugated-wire structure shows a negative refractive index in the plane *parallel* to the glass substrate (i.e., the xy-plane in Fig. 5.1(a)). The structure can be considered as an LCR circuit, where each pair of opposing wires represents the capacitive part. The inductance is provided by every corrugated wire, which simultaneously acts as a diluted metal. Note that the unit cell is centrosymmetric along the propagation direction and, thus, not bi-anisotropic. Furthermore, the unit cells look similar for light impinging from the x- or y-direction.

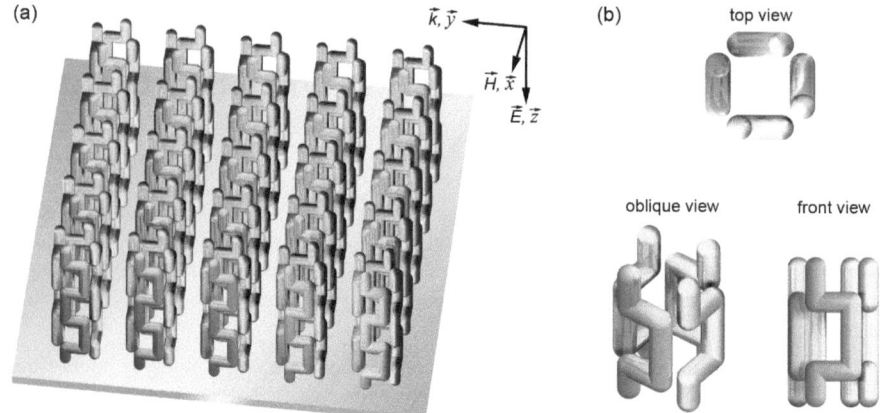

Figure 5.1: Artist's view of the corrugated-wire structure proposed in Ref. [145]. This metamaterial exhibits a negative refractive index in both planes parallel to the substrate. (a) Oblique-view illustration of the design and polarization of the incident electromagnetic field are illustrated. (b) Top view, oblique view, and front view of the unit cell.

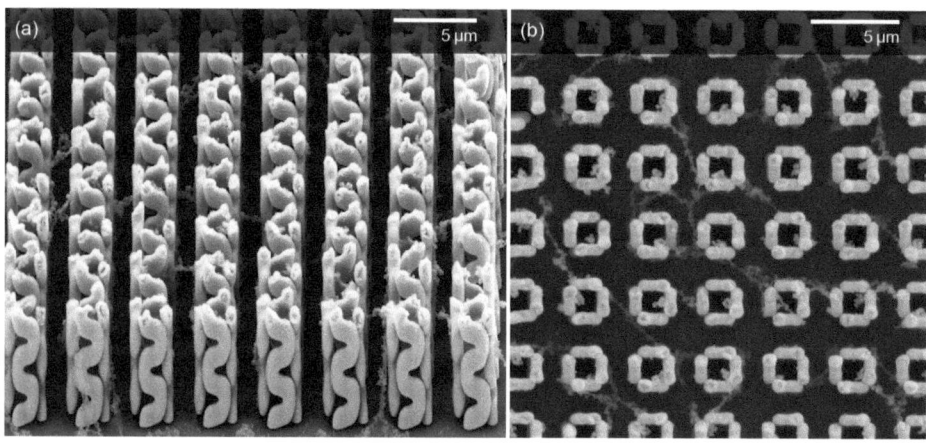

Figure 5.2: Experimental realization of a bulk metamaterial by using direct laser writing and electroplating of gold. (a) Oblique-view electron micrograph of the fabricated structure. The fluff around the metallic wires stems from the remaining photoresist which had not been removed during plasma etching. (b) Top-view electron micrograph.

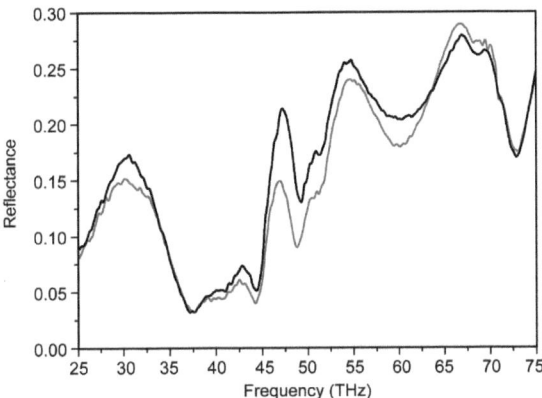

Figure 5.3: Measured reflectance spectrum under oblique incidence, i.e., the light beam is tilted by $20° \pm 5°$ to the normal z-axis. The gray and black curves relate to the configuration, where the wavevector has components pointing in x- and y-direction, respectively (see Fig. 5.1(a)). Within some limits, all features are reproduced in both configurations, i.e., the structure behaves nearly similar in both propagation directions.

Since the set of geometric parameters proposed in Ref. [145] is currently not producible, we have chosen to increase the size of the unit cell in order to obtain viable feature sizes. For the modified configuration, the structure is not expected to show a negative refractive index since the spectral regions where $\mathrm{Re}(\varepsilon) < 0$ and $\mathrm{Re}(\mu) < 0$ do not overlap anymore. Nevertheless, we intended to check whether it is, in principle, possible to fabricate this structure. Therefore, we processed the negative template into the positive-tone photoresist AZ 9260 *via* DLW and, subsequently, filled up the resulting holes by electroplating of gold [33]. For metallization, we applied an electric current of $0.235\,\mu\mathrm{A}$ (corresponding to a current density of $3 \cdot 10^{-3}\,\mathrm{A/cm^2}$), leading to a total growth time of about 45 minutes.

The corrugated-wire structure is mechanically not very stable due to the high aspect ratio of lateral and vertical dimensions. Moreover, it has a small footprint which makes the fabrication process fairly demanding. Especially, the small footprint causes severe problems since air inclusions prevent the holes from being filled to an equal level. These inclusions are formed during dipping the template into the aqueous electrolyte solution. We solved this problem by exchanging air by carbon dioxide (CO_2) in a self-made gas-tight chamber. In fact, volumes of gaseous carbon dioxide entirely dissolve in water and, thus, allow the electrolyte solution to get connected to the ITO cathode. As a result, even very thin and complex-shaped channels can be infiltrated. Moreover, AZ 9260 is very sensitive to dehydration. If the resist runs dry, numerous cracks are formed after development which can be mainly attributed to warpings due to the gold deposition. It turns out that introducing predetermined breaking points (e.g., walls around the functional domain) reduces the chance of damage considerably.

As shown in Fig. 5.2(a) and (b), we fabricated structures with two periods in z-direction and 15 periods in the xy-plane. The size of the unit cell was chosen to be $4\times4\times4\,\mu\mathrm{m}^3$. To get some information about the structure, we measured the spectrum by using the Bruker Tensor 27

Figure 5.4: Artist's view of a fully isotropic negative-index metamaterial proposed by Güney et al. The design proofs that the required connectivity of metallic elements does not preclude a negative refractive index. **(a)** Basic building block of the structure. **(b)** Arranging the unit cell from (a) in a cubic lattice results in a bulk isotropic metamaterial. Taken from [154].

FTIR spectrometer (see Fig. 5.3). The spectral position of the magnetic resonance is expected to be found in the mid-IR (at frequencies of around 45 THz). Since the glass substrate becomes opaque at frequencies below 50 THz, we confined ourselves to measure the reflectance only.

Due to geometrical restrictions of the microscope spectrometer, it is not possible to measure the spectrum of the metamaterial parallel to the substrate plane. Regrettably, this is the excitation geometry for which a negative refractive index has been proposed [145]. However, to come as close as possible to the desired configuration, we probed the structure under oblique incidence, i.e., the incident light beam is tilted by $20° \pm 5°$ to the normal z-axis. Moreover, we restricted the wavevector $\vec{k}$ to be either parallel to the xz- or the yz-plane (see Fig. 5.1(a)) which enables a partial excitation of in-plane resonances of interest.

In summary, we made an important step towards negative-index bulk metamaterials for the IR. Current restrictions related to attainable feature sizes are mainly rooted in the photoresist which we had to use out of specification. Unfortunately, no other suitable high-resolution positive-tone resists are commercially available at the moment. However, once it becomes available, there are no more conceptual problems. Thus, realizing appropriate bulk designs by using DLW and advanced coating processes is rather a matter of time. Since the experimental constraints have been defined, new compatible proposals have been reported recently like, e.g., the 3D-isotropic design shown in Fig. 5.4 [154].

6. Conclusions and Outlook

Since photonic metamaterials became an active field of research, planar composite nanostructures have played the key role in scientific investigations at near-infrared and visible frequencies. This is due to the fact that state-of-the-art nanolithography and deposition technologies mainly involve two-dimensional (2D) processes. However, the predicted effects of metamaterials are based on a continuous phase change of electromagnetic waves while passing through a bulk medium. Thus, a main objective has been the experimental implementation of 3D metamaterials which consist of multiple unit cells along the propagation direction.

First attempts to attack this problem used stacking of multiple planar layers *via* extended 2D processes [28, 30, 31]. However, for practical reasons, an inherently three-dimensional (3D) fabrication approach would be preferable for this task. In the course of this Thesis, we developed corresponding methods and realized 3D as well as bulk photonic metamaterials for the infrared spectral range. As metamaterials essentially consist of metallic unit cells, we require (i) a lithographic process providing a 3D nanoscaled backbone and (ii) a suitable metallization technology which offers the possibility to infiltrate even complex-shaped template structures.

For the sake of realizing the template structures, we utilized (i) direct laser writing (DLW) which is known to be a very flexible and versatile tool to fabricate 3D photonic crystals [8, 124]. Indeed, it can be considered as the rapid prototyping solution for the nanoscale. For many years, the accessible feature sizes were not sufficiently small to be convenient for realizing sub-wavelength unit cells required for near- or mid-infrared metamaterials. However, recent developments of this method in terms of automation and reproducibility have considerably softened these limitations. The working principle of DLW is based on two-photon absorption in a photoresist which occurs only within the high-intensity focal volume of a laser beam. The deposited energy induces a local chemical reaction changing the solubility of the resist. Scanning the focus of the laser beam relative to the sample defines the features of the structure. For negative-tone resists (e.g. SU-8), a following development process exposes the intended template, whereas a positive-tone resist (e.g. AZ 9260) yields the negative.

The 3D polymer template now serves as a framework for (ii) metallization. Notably, directed coating techniques like electro-beam evaporation or plasma sputtering are unsuitable for our purpose. These techniques would merely cover the outer surface leaving the interior uncoated. Hence, most of the template features would be non-functional with regard to a magnetic response. To avoid this problem, we have utilized methods which are either able to (ii.a) isotropically coat 3D templates or (ii.b) cast the respective negative structure.

In matters of the former approach, (ii.a) chemical vapor deposition (CVD) of silver has emerged to be an appropriate candidate [32]. We use the ligand-stabilized silver β-dikentonate precursor (1,5-cyclooctadiene)(1,1,1,5,5,5-hexafluoro-acetylacetonato)silver$^{(I)}$ [121] which is evaporated in an evacuated reaction chamber. The self-made CVD apparatus is fully automated enabling a cyclic deposition mode, i.e., sublimation and decomposition of the precursor substance are temporally separated. Emerging organic by-products are pumped out after each cycle to avoid contamination of the silver layer. Spectroscopic measurements have shown that the measured data could be nicely fitted by Drude parameters obtained from reference measurements [76, 77]. Slight deviations from the reference parameters arise from scattering due to surface roughness. However, the metal substructure has a negligible impact being by orders of magnitude smaller than typical feature sizes of polymer templates.

In matters of casting, (ii.b) gold electroplating based on an aqueous solution of sodium disulfioaurate[I] [33] has been established[1] which we use to gradually fill up the channels of a negative polymer structure. Therefore, the templates require a transparent conducting layer (e.g. indium tin oxide) beneath the photoresist which acts as an electrode during deposition. For structures with small voids, air inclusions might prevent the electrolyte solution from reaching this electrode which leads to unfilled gaps. We solved this problem by exchanging air by carbon dioxide in a self-made gas-tight chamber. Since carbon dioxide entirely dissolves in water, possible air inclusions disappear. As a result, even very thin and complex-shaped channels can be infiltrated. After electrodeposition, the polymer backbone is removed by using air plasma for several hours.

By using DLW in combination with silver CVD, we realized an array of upside-down 3D split-ring resonators (SRRs). After the metallization process, a corrugated metal surface results. Additional structuring *via* focused-ion beam milling leads to electrically separated and laterally connected SRRs. Measured and calculated spectra show a good agreement for all fabricated structures. Especially, the positions of the resonances in transmittance are nicely reproduced.

At first, the fabricated metamaterials have been examined with regard to the origin of appearing resonances. This was motivated by the fact that their spectral position considerably varied for certain geometries. Numerical calculations of the current densities for each case revealed that electrically connecting adjacent SRRs along the electric field component of the incident light tilts the orientation of the unit cell by 180°. Additionally, we realized that the fundamental magnetic resonance has shifted to the static limit. Hence, the experimentally observed resonances rather correspond to a higher mode.

To understand the underlying physics of the different metamaterial designs, we benefited from the advantage that the optical spectra and material parameters of the array of separated 3D SRRs can be calculated analytically in quasi-static approximation. Here, it was important to realize that the unit cells are not centrosymmetric along the propagation direction of the incident light. Therefore, the electric component of light induces both a parallel electric and a perpendicular magnetic dipole moment. A related excitation geometry is also found for the magnetic component of light. This cross-coupling phenomenon is associated with bi-anisotropy. To confirm the results of the analytic model, we determined the electric, magnetic, and cross-coupling response (given by the permittivity ε, the permeability μ, and the bi-anisotropy parameter ξ, respectively) of the material to electromagnetic excitations by post-processing of our numerically calculated spectra. For this purpose, we developed a parameter retrieval based on an inversion of the respective Fresnel equations. Applying this procedure yields simultaneous negative real parts of ε and μ for all configurations. In the case of isotropic metamaterials, this

---

[1]The electroplating process has been set up by Justyna K. Gansel at the Institut für Angewandte Physik (Karlsruhe Institute of Technology).

would be a sufficient condition for obtaining a negative refractive index, i.e., $n<0$. In contrast, for our bi-anisotropic media, the effect of cross-coupling rather suppresses a negative refractive index.

However, from general theoretical considerations we know that bi-anisotropy does not inherently forbid negative-index metamaterials. Hence, we have designed a bi-anisotropic metamaterial with an enhanced electric response. For fabrication, we utilized DLW and electron-beam shadow evaporation of silver to fabricate an extended version of the 3D SRR array [110], i.e., we added elevated metal wires parallel to the electric field vector of the incident light. These wires act as a diluted metal and provide an electric resonance at a designed spectral region. As expected, this composite structure showed a similar magnetic and electric response leading to negative values for ε and μ. Beyond that, we demonstrated a negative refractive index from 77 THz to 83 THz (i.e., for wavelengths between 3.6 μm and 3.9 μm). Here, the bi-anisotropic behavior even actively supports this feature in that it broadens the spectral range where $n<0$.

Obviously, the 3D SRR array and its descendants cannot be considered as bulk metamaterials, but they clearly represent an important first step towards this direction. Recently, first design proposals accounting for the experimental constraints of our fabrication method have been reported [145, 154]. Although the proposed geometrical parameters of these structures are not feasible for fabrication to date, we realized an enlarged version of a corrugated-wire structure [145] by using DLW and electroplating of gold. Clearly, we have overcome conceptual obstacles in fabrication and opened the door for future developments. Besides that, we exposed the intrinsic constraints of 3D nanofabrication enabling a targeted development of new metamaterial designs.

Successive progresses in both design and fabrication of bulk composite materials will certainly trigger further breakthroughs in future. Especially, the unique characteristics of metamaterials contain a lot of potential for real-world applications [16, 33, 155]. Of course, there is always some room for improvement. Related to our fabrication process, further developments concerning surface functionalization and feature size reduction are required. The former issue is urged on providing as many degrees of freedom as possible in view of structure designs. Especially, a solution for the substrate metallization—naturally arising from isotropic coating processes—must be provided. The feature size reduction, however, is driven by the motivation to deliver new materials for telecommunication technology. Recently, first promising attempts in that direction have been reported in the context of STED-DLW [156].

A. Background and Details

A.1. Wood Anomaly

Wood anomalies [157, 158] were first observed in the spectrum of light resolved by optical diffraction gratings. They appear as rapid variations in the intensity of various diffracted spectral orders in certain narrow frequency bands. In 1902, R. W. Wood discovered these effects in experiments on reflection gratings, and called them "anomalies" since they could not be explained by ordinary grating theory.

The variations in the intensity are caused by (i) Rayleigh diffraction which occurs for any polarization of the incident light wave and (ii) resonant coupling to surface-plasmon modes if the polarization of the incident light is perpendicular to a metal grating. For an instructive derivation of case (i) (that is also called Rayleigh-Wood anomaly), we consider a nanostructure consisting of a pattern with periodicity a in x-direction. The structure itself is carried by a glass substrate having a refractive index of $n_{\text{glass}} = 1.5$. This corresponds to a typical situation of measuring the transmittance of our fabricated samples.

If light is impinging from the vacuum half-space ($n_{\text{vac}} = 1$), the incident wave vector reads

$$\vec{k}_{\text{in,vac}} = \begin{pmatrix} k_x \\ k_y \\ k_z \end{pmatrix}.$$

Due to the periodicity in x-direction, we can add an even multiple N to the reciprocal lattice vector $(2\pi)/a$ in k-space. If also energy conservation

$$\left|\vec{k}_{\text{out}}^N\right| = n_{\text{glass}} k = n_{\text{vac}} k$$

is taken into account, the wave vector of the transmitted light which is diffracted into the glass substrate will be given by

$$\vec{k}_{\text{out,glass}}^N = \begin{pmatrix} k_x + \frac{2\pi N}{a} \\ k_y \\ \pm\sqrt{k^2 n_{\text{glass}}^2 - k_y^2 - (k_x + \frac{2\pi N}{a})^2} \end{pmatrix}.$$

Hence, for normal incidence ($k_x = k_y = 0$) the first diffraction order appears at $\lambda_{\text{glass},1} = 1.5\,a$ in glass and $\lambda_{\text{vac},1} = a$ in vacuum. At these wavelengths a dip in transmittance is observable and for $\lambda > \lambda_{\text{glass},1}$ the structure must be treated as a photonic crystal.

A.2. Reciprocity in Optics

The reciprocity theorem (in terms of H. Lorentz) states that the relation between an oscillating current and an induced electric field is unchanged if one exchanges the points where the current is placed and the field is measured. Suppose that we place a local current density $\vec{j}_1$ inside the medium at a defined position "1". The current induces an electric field $\vec{E}_2$ and a magnetic field $\vec{H}_2$ at another position "2". Similarly, we place a second local current density $\vec{j}_2$ at position "2" which, analogously, generates the fields $\vec{E}_1$ and $\vec{H}_1$. All vectors are supposed to be harmonic functions of time including a frequency ω. The medium is considered to be reciprocal if the relation

$$\int \vec{j}_1 \cdot \vec{E}_2 \, \mathrm{d}V = \int \vec{j}_2 \cdot \vec{E}_1 \, \mathrm{d}V \tag{A.1}$$

holds. Notably, for this definition, there must *not* exist any external sources that emit waves impinging from infinitely far away.

Reciprocity introduces a time-reversal symmetry to the material system and, thus, simplifies the appearance of the material parameters. As a consequence, some optical measures like the refractive index and the transmittance are equal for both anti-parallel propagation directions.

For the context of this Thesis, it might be interesting to find formal conditions for material parameters which describe a reciprocal bi-anisotropic medium. We start according to Ref. [35] by setting

$$\vec{H}_2 \cdot (\nabla \times \vec{E}_1) - \vec{E}_1 \cdot (\nabla \times \vec{H}_2) + \vec{E}_2 \cdot (\nabla \times \vec{H}_1) - \vec{H}_1 \cdot (\nabla \times \vec{E}_2) \tag{A.2}$$

$$\stackrel{(2.15)-(2.16)}{=} \vec{H}_2 \cdot (-i\omega \vec{B}_1) - \vec{E}_1 \cdot (\vec{j}_2 + i\omega \vec{D}_2) + \vec{E}_2 \cdot (\vec{j}_1 + i\omega \vec{D}_1) - \vec{H}_1 \cdot (-i\omega \vec{B}_2)$$

$$\stackrel{(2.5)-(2.10)}{=} i\omega\mu_0 \underbrace{\left(\vec{H}_1 \underline{\mu} \vec{H}_2 - \vec{H}_2 \underline{\mu} \vec{H}_1 \right)}_{\text{(I)}} + i\omega\varepsilon_0 \underbrace{\left(\vec{E}_2 \underline{\varepsilon} \vec{E}_1 - \vec{E}_1 \underline{\varepsilon} \vec{E}_2 \right)}_{\text{(II)}}$$

$$+ \frac{i\omega}{c_0} \left(\underbrace{\vec{E}_2 \underline{\xi} \vec{H}_1 + \vec{H}_1 \underline{\zeta} \vec{E}_2}_{\text{(III)}} \underbrace{- \vec{E}_1 \underline{\xi} \vec{H}_2 - \vec{H}_2 \underline{\zeta} \vec{E}_1}_{\text{(IV)}} \right)$$

$$+ \underbrace{\vec{E}_2 \vec{j}_1 - \vec{E}_1 \vec{j}_2}_{\text{(V)}} \;.$$

From (A.1), the reciprocity theorem can be derived only if the ansatz (A.2) and, additionally, the terms (I)–(IV) equal zero. If this condition is fulfilled, (A.1) can be obtained by integration over the medium's volume.

At first, the ansatz of (A.2) can be transformed by vector analysis [34] to

$$\nabla \cdot \left(\vec{E}_1 \times \vec{H}_2 - \vec{E}_2 \times \vec{H}_1 \right) \;.$$

Integrating over the whole volume of the medium and using the divergence theorem leads to

$$\int \nabla \cdot \left(\vec{E}_1 \times \vec{H}_2 - \vec{E}_2 \times \vec{H}_1 \right) \mathrm{d}V = \oint_S \left(\vec{E}_1 \times \vec{H}_2 - \vec{E}_2 \times \vec{H}_1 \right) \mathrm{d}A \;.$$

By assuming a lossy medium, the fields decay exponentially with distance from the localized currents. Thus, in the case of large material sizes, the surface integral vanishes.

Next, we have to find conditions for which (I) and (II) equal zero. Having a closer look at these terms reveals that cancelation merely results for symmetric permittivity and permeability tensors whose transposition equals their pristine version, i.e., $\underline{\varepsilon} = \underline{\varepsilon}^t$ and $\underline{\mu} = \underline{\mu}^t$. By way of example, inserting asymmetrical off-diagonal entries to $\underline{\varepsilon}$, results in non-zero dot products of $\vec{E}_1$- and $\vec{E}_2$-field components.

A related situation is given for (III) and (IV). Here, the summands cancel either if $\underline{\xi} = -\underline{\zeta}^t$ or $\underline{\zeta} = -\underline{\xi}^t$. Inserting, e.g, $\underline{\zeta} \equiv +\underline{\xi}^t$, yields non-vanishing dot products of $\vec{E}_{1,2}$- and $\vec{H}_{2,1}$-field components.

From the upper discussion, it becomes clear that isotropic media are reciprocal by definition. For bi-anisotropic and bi-isotropic media, the field configuration might become more difficult to understand since cross-term parameters allow to change the spatial direction of electric and magnetic fields in an arbitrary manner. Thus, many controversial discussions have been devoted to the existence of non-reciprocal bi-isotropic and bi-anisotropic media [159–165].

A well-known non-reciprocal optical system is the "Faraday isolator". Here, the time-inversion symmetry of light propagation is broken due to the preference of a spatial direction defined by an external *static* magnetic field.

A.3. Lorentz Oscillator Model

The Lorentz oscillator model assumes that electrons are held in a stable orbit with respect to the nucleus. A "spring" represents the restoring force for small displacements from the equilibrium position. The negatively charged electron and the positively charged nuclei form electric dipoles with a magnitude proportional to their separation. Additionally, damping is included in the model to account for energy losses due to collisional processes. In solids, e.g., this would typically occur through interactions of electrons and phonons in the crystal [2,3].

The equation of motion of an electron in a harmonic potential which is excited by an electromagnetic wave reads

$$m_\mathrm{e} \frac{\partial^2 \vec{r}_0}{\partial t^2} + m_\mathrm{e} \gamma \frac{\partial \vec{r}_0}{\partial t} + m_\mathrm{e} \omega_0^2 \vec{r}_0 = -e\vec{E}_0 \; \mathrm{e}^{i(\vec{k}\vec{r}_0 - \omega t)} \; , \tag{A.3}$$

where e denotes the electron charge, γ the damping factor, and ω_0 the eigenfrequency of a bound electron. In (A.3), we used the electron mass although the reduced mass $m_\mathrm{red}^{-1} = m_\mathrm{e}^{-1} + m_\mathrm{nucleus}^{-1}$ would be more convenient. However, since $m_\mathrm{e} \ll m_\mathrm{nucleus}$, we may safely take $m_\mathrm{red} \approx m_\mathrm{e}$ here.

If the velocity of the electron is non-relativistic, i.e., $v \ll c_0$, the magnetic force can be neglected in the calculation. Remarkably, only the real part of (A.3) leads to a physically relevant solution. In the following discussion, we assume that the incident wavelength λ of light is much larger than the Bohr radius, i.e.,

$$\mathrm{e}^{i\vec{k}\vec{r}_0} = 1 + \mathcal{O}\left(\frac{r_0}{\lambda}\right) \approx 1 \; .$$

As (A.3) is a linear, non-homogeneous differential equation, the solution is separable to a homogeneous and a particular part. For $t \gg 1/\gamma$, the homogeneous solution decays. Using

$$\vec{r}_{0,\mathrm{part}}(t) = \vec{A}\, \mathrm{e}^{-i\omega t}$$

as an ansatz for the particular solution of the forced oscillation results in

$$\left(-\omega^2 - i\gamma\omega + \omega_0^2\right) \vec{A} = -\frac{e}{m_e} \vec{E}_0 \ .$$

Hence, the dipole moment of the electron reads

$$\vec{p}(t) = -e\vec{r}_0(t) = -e\vec{A}\,e^{-i\omega t} = \left(\frac{e^2}{m_e(\omega_0^2 - \omega^2 - i\gamma\omega)}\right) \vec{E}_0\,e^{-i\omega t} = \alpha_e\,e^{-i\omega t} \ , \qquad (A.4)$$

where α_e is the complex electric polarizability. With the definition of the permittivity

$$\varepsilon(\vec{r},\omega) = 1 + \chi_e = 1 + \frac{n_0(\vec{r})\,\alpha_e(\omega)}{\varepsilon_0}$$

and by summing over all oscillators in the solid (i.e., transitions of bound electrons as well as vibrational bands), we finally obtain

$$\varepsilon(\omega) = 1 + \frac{n_0(\vec{r})e^2}{m_e\varepsilon_0} \sum_j \frac{f_j}{\omega_j^2 - \omega^2 - i\gamma_j\omega} \ . \qquad (A.5)$$

Here, n_0 indicates the density of atoms, χ_e the electric susceptibility, and f_j the number of electrons which have the same damping and eigenfrequency (i.e., the oscillator strength). Experimental data shows that the absorption strength of the oscillator actually varies considerably between different atomic transitions. With the benefit of hindsight, we know that this is caused by the variation of the quantum mechanical transition probability. In classical physics, however, there is no explanation for this phenomenon. Hence, we assign an oscillator strength to each transition such that $\sum_j f_j = 1$.

For the special case of free electrons which do not feel any restoring force, the term of (A.3) proportional to ω_0^2 vanishes. Hence, the eigenfrequency ω_j becomes zero and the equation of $\varepsilon(\omega)$ simplifies to (2.47). Note that we replaced the electron mass m_e by the effective electron mass m_{eff} in (2.47) to account for the contribution of the periodic potential of the crystal lattice.

A.4. Mathematical Proof that (2.39) Solves (2.38)

In order to fulfill thermodynamical laws, the real part of a material's impedance must be greater than zero. We constitute this statement by using the model of a hypothetical current sheet (shown in Fig. A.1) located at position $x = x_0$ which radiates into a medium. The current sheet is assumed to be uniform and infinite in the y- and z-directions [74]. The wave equation of this configuration is derived from the Maxwell equations (2.13)–(2.16)

$$\begin{aligned}
\nabla \times (\nabla \times \vec{E}(\vec{r},t)) = -\Delta \vec{E} &= i\omega\mu_0\mu(\nabla \times \vec{H}(\vec{r},t)) \\
&= \frac{\omega^2 n^2}{c_0^2} \vec{E}(\vec{x},t) + i\omega\mu_0\mu\,\vec{j}(t)\,\delta(\vec{r} - \vec{r}_0) \qquad (A.6)
\end{aligned}$$

Here, the Dirac delta function is used in order to restrict the current flow to the thin current sheet. The current density is explicitly given by $\vec{j}(t) = \vec{j}_0\,e^{-i\omega t}$. Without loss of generality, we

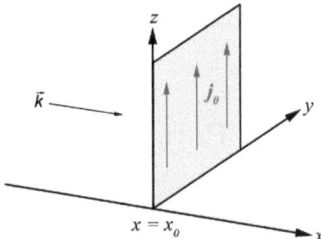

Figure A.1: A current sheet located at position $x=x_0$, is excited by an external monochromatic harmonic wave. The current sheet is assumed to be uniform and infinite in the y- and z-directions. Adapted from Ref. [74].

are allowed to choose the coordinate system in a way that the wave propagates in x-direction. Hence, the vectors in (A.6) can be replaced by scalars.

As an ansatz for the inhomogeneous differential equation we choose

$$\begin{aligned} E(x,t) &= -\frac{1}{2} Z(\omega) j_0 \, e^{i(nk|x-x_0|-\omega t)} \\ &= -\frac{c_0 \mu_0 \mu}{2n} j_0 \, e^{i(nk|x-x_0|-\omega t)} \,, \end{aligned} \quad (A.7)$$

where the second spatial derivation reads

$$\frac{\partial^2 E(x,t)}{\partial x^2} = -\frac{c_0 \mu_0 \mu}{2n} j_0 \, ink \, e^{ink|x-x_0|-\omega t} \left(\frac{\partial^2 |x-x_0|}{\partial x^2} + ink \left(\frac{\partial |x-x_0|}{\partial x} \right)^2 \right).$$

Notably, $\frac{\partial}{\partial x}|x-x_0| = 2\Theta(x-x_0) - 1$ and $\frac{\partial^2}{\partial x^2}|x-x_0| = 2\delta(x-x_0)$, where we use the Dirac delta function $\delta(x)$ and the Heaviside function $\Theta(x)$. Thus, the differential equation results in

$$\underbrace{-2ikc_0 \, \delta(x-x_0)}_{\text{(I)}} + \underbrace{nk^2 c_0 \left(2\Theta(x-x_0) - 1\right)^2}_{\text{(II)}} - \underbrace{n\omega^2/c_0}_{\text{(III)}} = \underbrace{-2i\omega \, \delta(x-x_0) \, e^{-ink|x-x_0|}}_{\text{(IV)}}.$$

For $x > x_0$ and $x < x_0$, the Dirac delta functions in (I) and (IV) are strictly zero. Since (II) equals (III) in both cases, the differential equation is solved.

For $x = x_0$, one has to integrate over an infinitely small distance $(x_0 \pm \kappa)$ so that

(I): $\displaystyle\int_{x_0-\kappa}^{x_0+\kappa} -2ikc_0 \, \delta(x-x_0) \, \mathrm{d}x \stackrel{\kappa \to 0}{=} -2ikc_0 \,,$

(II): $\displaystyle\int_{x_0-\kappa}^{x_0+\kappa} nk^2 c_0 \left(2\Theta(x-x_0) - 1\right)^2 \mathrm{d}x = nk^2 c_0 \left[|x-x_0|\right]_{x_0-\kappa}^{x_0+\kappa} = 0 \,,$

(III): $\displaystyle\int_{x_0-\kappa}^{x_0+\kappa} \frac{n\omega^2}{c_0} \, \mathrm{d}x = \frac{n\omega^2}{c_0} \left[x\right]_{x_0-\kappa}^{x_0+\kappa} \stackrel{\kappa \to 0}{=} 0 \,,$

(IV): $\displaystyle\int_{x_0-\kappa}^{x_0+\kappa} -2i\omega \, \delta(x-x_0) \, e^{-ink|x-x_0|} \, \mathrm{d}x \stackrel{\kappa \to 0}{=} -2i\omega \, e^{-ink|x_0-x_0|} = -2i\omega \,.$

In summary, we find that (II) = (III) = 0 and (I) = (IV), i.e., the differential equation is again solved.

A.5. Fresnel Equations of Purely Bi-Anisotropic Media

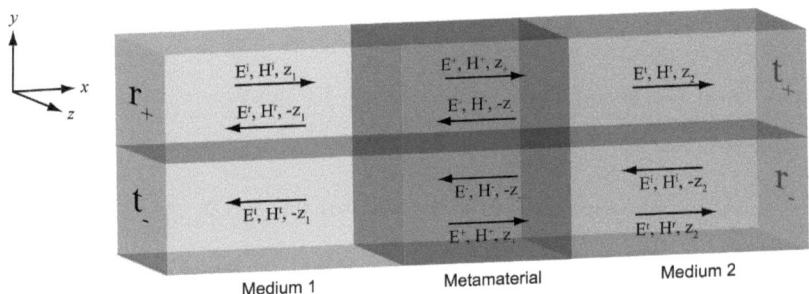

Figure A.2: Illustration of the field components for a generalized version of Fresnel's equations used to retrieve the effective parameters for purely bi-anisotropic media. The metamaterial (dark) is clad between two isotropic media 1 and 2 (e.g., air and glass). Adapted from Ref. [51].

Inverting the Fresnel equations allows us to determine the constitutive material parameters from the complex transmission and reflection coefficients. In the style of Ref. [51], the retrieval procedure is derived for purely bi-anisotropic media, where the induced field components include an angle $\varphi = 90°$ with the vectors which excited them. Since this correlates with the discussion in section 2.8.1, we will use the same nomenclature as previously.

A monochromatic and linearly polarized electromagnetic wave impinges under normal incidence from an isotropic material of relative impedance $z_1 = Z_1/Z_0$ (e.g., air or vacuum) onto a bi-anisotropic metamaterial slab of thickness d_s. The related fields are denoted by E^i and H^i. After passing the metamaterial, the wave is transmitted into another isotropic material of relative impedance $z_2 = Z_2/Z_0$ (e.g., a glass substrate). The geometry and the nomenclature used in our discussion are illustrated in Fig. A.2. Considering the constitutive relations of a bi-anisotropic material (2.48)–(2.49) and introducing the plane-wave ansatz $E^\pm$ and $H^\pm$ for both propagation directions ($\pm$) into Maxwell's equations immediately leads to linear eigensolutions as long as the wave propagates along this axis. A change in polarization could occur for oblique incidence of light onto the slab and / or for chiral media.

The bulk impedance of the bi-anisotropic material is $Z_+ = E^+/H^+$ for propagation in the (+)-direction and $-Z_- = -E^-/H^-$ for propagation in the (−)-direction. In analogy to (2.52), the relative impedances are given by

$$z_\pm := \frac{Z_\pm}{Z_0} = \frac{\mu_{zz}}{\pm n_{ba} - i\xi_{yz}}, \qquad (A.8)$$

where we denote the vacuum impedance as Z_0. Notably, $z_+ \neq -z_-$.

Next, we assume that the boundary conditions of the tangential components of E and H are continuous and that $HZ_0 = E/z_i$. Moreover, we introduce the complex reflection and transmission coefficients for a wave impinging from the (+)-direction, i.e., $r_+ = E^r/E^i$ and $t_+ = E^t/E^i$, respectively. Hence, we obtain at $x=0$

$$(1+r_+)E^i = E^+ + E^-, \tag{A.9}$$

$$\left(\frac{(1-r_+)E^i}{z_1}\right) = \left(\frac{E^+}{z_+}\right) + \left(\frac{E^-}{z_-}\right) \tag{A.10}$$

and at $x = d_s$

$$E^+ e^{in_{ba}k_0 d_s} + E^- e^{-in_{ba}k_0 d_s} = t_+ E^i, \tag{A.11}$$

$$\left(\frac{E^+ e^{in_{ba}k_0 d_s}}{z_+}\right) + \left(\frac{E^- e^{-in_{ba}k_0 d_s}}{z_-}\right) = \frac{t_+ E^i}{z_2}. \tag{A.12}$$

With (A.9)–(A.10) ((A.11)–(A.12)) we express E^+/E^i and E^-/E^i as linear functions of r_+ (t_+):

$$\frac{E^+}{E^i} = a_+ + b_+ r_+ \quad \text{and} \quad \frac{E^+}{E^i} = c_+ + d_+ t_+,$$

$$\frac{E^-}{E^i} = a_- + b_- r_+ \quad \text{and} \quad \frac{E^-}{E^i} = c_- + d_- t_+.$$

This yields two linear relationships between r_+ and t_+, i.e.,

$$t_+ = \alpha + \beta r_+ \quad \text{and}, \tag{A.13}$$
$$t_+ = \gamma + \delta r_+, \tag{A.14}$$

where

$$\alpha = e^{in_{ba}k_0 d_s}\left(\frac{1 - z_-/z_1}{1 - z_-/z_2}\right),$$

$$\beta = e^{in_{ba}k_0 d_s}\left(\frac{1 + z_-/z_1}{1 - z_-/z_2}\right),$$

$$\gamma = e^{-in_{ba}k_0 d_s}\left(\frac{1 - z_+/z_1}{1 - z_+/z_2}\right),$$

$$\delta = e^{-in_{ba}k_0 d_s}\left(\frac{1 + z_+/z_1}{1 - z_+/z_2}\right).$$

We want to determine the three complex material parameters ε_{yy}, μ_{zz} and ξ_{yz}, which directly depend on n_{ba}, z_+ and z_-, from the complex transmittance and reflectance of the material. Therefore, (A.13) and (A.14) alone are not sufficient to solve the problem. Additionally, we need to consider the case of propagation in the (−)-direction as well. In this case, (A.9)–(A.12) take a similar form as previously, except that we have to substitute (see Fig. A.2)

(+)-direction:	z_1	z_2	z_+	z_-
	⇓	⇓	⇓	⇓
(−)-direction:	$-z_2$	$-z_1$	z_-	z_+

Consequently, we obtain the following equations corresponding to (A.13) and (A.14) for the (−)-direction, i.e.,

$$t_- = \alpha' + \beta' r_- \,, \tag{A.15}$$
$$t_- = \gamma' + \delta' r_- \,, \tag{A.16}$$

where

$$\alpha' = e^{in_{ba}k_0 d_s} \left(\frac{1 + z_+/z_2}{1 + z_+/z_1} \right),$$

$$\beta' = e^{in_{ba}k_0 d_s} \left(\frac{1 - z_+/z_2}{1 + z_+/z_1} \right),$$

$$\gamma' = e^{-in_{ba}k_0 d_s} \left(\frac{1 + z_-/z_2}{1 + z_-/z_1} \right),$$

$$\delta' = e^{-in_{ba}k_0 d_s} \left(\frac{1 - z_-/z_2}{1 + z_-/z_1} \right).$$

Notably, $t_+/z_2 = t_-/z_1$ (calculation not detailed here) which results in $T = T_+ = T_-$, i.e., the transmittance T does not depend on the side from which light impinges onto the slab.

We now need to invert (A.13)–(A.16) in order to calculate z_+, z_- and n_{ba} for known t_+, r_+, t_- and r_-. Multiplying (A.13) by (A.16) and (A.14) by (A.15) leads to

$$t_+ t_- = \alpha \gamma' + \beta \gamma' r_+ + \alpha \delta' r_- + \beta \delta' r_+ r_- \,, \tag{A.17}$$
$$t_+ t_- = \gamma \alpha' + \delta \alpha' r_+ + \gamma \beta' r_- + \delta \beta' r_+ r_- \,, \tag{A.18}$$

whereas

$$\alpha \gamma' = \frac{(1 - z_-/z_1)(1 + z_-/z_2)}{(1 + z_-/z_1)(1 - z_-/z_2)},$$

$$\gamma \alpha' = \frac{(1 - z_+/z_1)(1 + z_+/z_2)}{(1 + z_+/z_1)(1 - z_+/z_2)},$$

$$\beta \gamma' = \frac{1 + z_-/z_2}{1 - z_-/z_2},$$

$$\delta \alpha' = \frac{1 + z_+/z_2}{1 - z_+/z_2},$$

$$\alpha \delta' = \frac{1 - z_-/z_1}{1 + z_-/z_1},$$

$$\gamma \beta' = \frac{1 - z_+/z_1}{1 + z_+/z_1},$$

$$\beta \delta' = 1,$$

$$\delta \beta' = 1.$$

It follows that (A.17) and (A.18) are similar equations for z_+ and z_- which can be rewritten as a second degree polynomial equation for $z_\pm$: $az_\pm^2 + bz_\pm + c = 0$ [32, 51]:

$$z_\pm = \frac{-b \mp \sqrt{b^2 - 4ac}}{2a} \tag{A.19}$$

with

$$a = t_+t_- - (1-r_+)(1-r_-),$$
$$b = (z_1 - z_2)(t_+t_- + 1 - r_+r_-) + (z_1 + z_2)(r_+ - r_-),$$
$$c = z_1 z_2 \left[-t_+t_- + (1+r_+)(1+r_-)\right].$$

Related to the results of section 2.6.1, the sign in (A.19) must be chosen such that the real part of the passive-medium impedance is positive definite. As already noted, z_+ is the relative impedance of the bi-anisotropic medium in the $(+)$-direction, and z_- is the opposite of the relative impedance in the $(-)$-direction which yields $\mathrm{Re}(z_+) > 0$ and $\mathrm{Re}(-z_-) > 0$, respectively.

To derive the refractive index, (A.13) and (A.14) are rewritten as

$$t_+ = e^{in_{\mathrm{ba}}k_0 d_{\mathrm{s}}} \left(\frac{1 + r_+ - (1-r_+)z_-/z_1}{1 - z_-/z_2} \right),$$

$$t_+ = e^{-in_{\mathrm{ba}}k_0 d_{\mathrm{s}}} \left(\frac{1 + r_+ - (1-r_+)z_+/z_1}{1 - z_+/z_2} \right).$$

Finally, we get an implicit expression for the (complex) refractive index n_{ba} [32, 51]

$$\cos(n_{\mathrm{ba}}k_0 d_{\mathrm{s}}) = \frac{t_+}{2} \left(\frac{1 - z_-/z_2}{1 + r_+ - (1-r_+)z_-/z_1} + \frac{1 - z_+/z_2}{1 + r_+ - (1-r_+)z_+/z_1} \right). \quad (\mathrm{A.20})$$

Notably, (A.20) has infinitely many solutions for n_{ba} due to the different branches of the inverse cosine. To choose the correct one, we proceed as proposed for the parameter retrieval for structures with inversion symmetry [7].

Once $z_\pm$ and n_{ba} are at hand, we deduce ε_{yy}, μ_{zz} and ξ_{yz} by using (A.19) and (A.20). The required relations, which express the material parameters in terms of the impedances and the refractive index, can be derived from (A.8) and (2.54) as

$$\varepsilon_{yy} = \frac{n_{\mathrm{ba}} + i\xi_{yz}}{z_+},$$
$$\mu_{zz} = z_+ (n_{\mathrm{ba}} - i\xi_{yz}),$$
$$\xi_{yz} = in_{\mathrm{ba}} \left(\frac{z_- + z_+}{z_- - z_+} \right).$$

A.6. Numerical Time-Domain Calculations

Analytic solutions of Maxwell equations for complex problems are often very difficult and mostly impossible to obtain. Hence, numerical routines have been developed which solve given electromagnetic problems either in frequency or time domain. At first view, it seems to be needless to differentiate between Maxwell equations in frequency and time domain as they can be easily converted by analytic Fourier transforms. However, on a numerical level the approaches to obtain stable and convergent solutions are fundamentally different.

In this section, we will rather focus on the principles of time-domain solvers. Here, we further distinguish between numerical solvers which use the differential or the integral form of

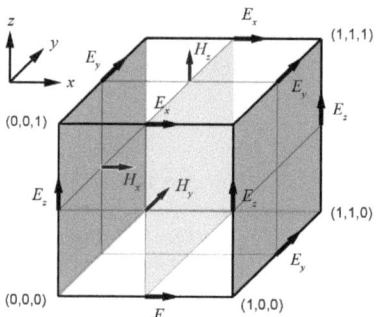

Figure A.3: Illustration of the Yee grid, i.e., a Cartesian grid system which is normally used for finite-difference time-domain calculations. The graph shows the positions of various field components. The $\vec{E}$-field components are placed in the middle of the edges and the $\vec{H}$-field components are in the center of the faces. Adapted from Ref. [139].

Maxwell's equations. During the last decades, several approaches have been proposed to solve the differential equations, i.e., the finite-difference time-domain (FDTD) approach [139, 166], the discontinuous Galerkin time-domain approach (DGTD) [167, 168], and the Krylov-subspace method [169].

The FDTD method, being the most popular approach, models the propagation of an electromagnetic wave in a spatial volume containing dielectric and / or metallic elements which represent the structure in question. At each time step, i.e., at each implementation of a finite-difference analog of Maxwell's curl equations at each spatial element, the incident wave is tracked as it first propagates to the structure and then interacts with it *via* surface-current excitation, diffusion, penetration, and diffraction. The main advantage of this approach is its simplification by analyzing the interaction at an instant time. Hence, it is not intended to solve the entire problem within one single step.

Time-stepping is accomplished by an explicit finite-difference procedure proposed in Ref. [139], where the electric and magnetic field components are placed in a Cartesian space grid (shown in Fig. A.3), i.e., the so-called "Yee grid". Here, each $\vec{E}$-field vector component is located midway between a pair of $\vec{H}$-field vector components, and *vice versa*. The evolution of the fields is now calculated at alternating half-time steps in a "leap-frog" manner. To understand the idea, we notice that at any point in space, the updated value of the $\vec{E}$-field ($\vec{H}$-field) in time is dependent on the stored value of the $\vec{E}$-field ($\vec{H}$-field) and the numerical curl of the local distribution of the $\vec{H}$-field ($\vec{E}$-field) in space. Transferring this fact to the Yee grid means that the $\vec{E}$-field and $\vec{H}$-field updates are translated so that $\vec{E}$-field updates are executed midway during each time step between successive $\vec{H}$-field updates. This time-stepping scheme avoids to solve simultaneous equations at once and yields lossless numerical wave propagation.

To provide an explicit example of how the algorithm works, we consider the Fourier-transformed (time-dependant) version of Maxwell's curl equation (2.15) in x-direction which is given by

$$\frac{\partial E_z}{\partial y} - \frac{\partial E_y}{\partial z} = -\frac{\partial B_x}{\partial t}.$$

Without loss of generality, we define $F(x,y,z,t) = F(0,0,0,t)$ to be a function at the point $(0,0,0)$ in the discretized Cartesian grid space, shown in Fig. A.3, evolved at a certain time step $t > 0$. Thus, the related finite-difference equation reads

$$\frac{E_z(0,1,\frac{1}{2},t) - E_z(0,0,\frac{1}{2},t)}{\Delta y} - \frac{E_y(0,\frac{1}{2},1,t) - E_y(0,\frac{1}{2},0,t)}{\Delta z}$$
$$= -\frac{B_x(0,\frac{1}{2},\frac{1}{2},t+\frac{1}{2}) - B_x(0,\frac{1}{2},\frac{1}{2},t-\frac{1}{2})}{\Delta t} \,. \tag{A.21}$$

For the y and z-direction, the finite-difference equations are set up in an analog manner. Note that all appearing field components are situated in the dark shaded plane (see Fig. A.3). Applying the same concept to the Fourier-transformed version of (2.15), i.e.,

$$\frac{\partial H_z}{\partial y} - \frac{\partial H_y}{\partial z} = \frac{\partial D_x}{\partial t} + j_x \,,$$

results in

$$\frac{H_z(\frac{1}{2},\frac{1}{2},0,t-\frac{1}{2}) - H_z(\frac{1}{2},-\frac{1}{2},0,t-\frac{1}{2})}{\Delta y} - \frac{H_y(\frac{1}{2},0,\frac{1}{2},t-\frac{1}{2}) - H_y(\frac{1}{2},0,-\frac{1}{2},t-\frac{1}{2})}{\Delta z}$$
$$= \frac{D_x(\frac{1}{2},0,0,t) - D_x(\frac{1}{2},0,0,t-1)}{\Delta t} - j_x\left(\frac{1}{2},0,0,t-\frac{1}{2}\right) \,. \tag{A.22}$$

In (A.22), all field components are placed in the light shaded plane, i.e., between two successive numeric evaluations of (A.21).

In contrast to finite-difference time-domain (FDTD) algorithms, the finite-integration technique (FIT) [137, 138] solves the integral rather than the differential Maxwell equations. Hence, the resulting quantities are rather the grid voltages and fluxes than the field components. The conservation of charges and energy is, therefore, inherently included which improves the solver's convergence and stability. Moreover, the FIT is also applicable for frequency-domain simulations.

If we use the same Yee-type grid as for the FDTD method, the FIT matrix equations can be transformed one-to-one and onto respective FDTD matrices. But cuboid meshes often cannot approximate the considered structure geometry as good as required. Precisely, the spatial discretization of curved structure features leads to "staircasing" effects. Especially for metallic features, unintended field enhancements at sharp edges lead to non-converging or falsified solutions. Hence, a lot of effort has been spent to extend time-domain calculations to non-orthogonal meshes (e.g., triangular or cylindric meshes) [168, 170–172].

Alternatively, one can also mimic curved boundaries of complex-shaped geometries *via* "Perfect Boundary Approximation™" [173]. In this case, the advantages of the Cartesian Yee grid is still given (i.e., fast calculation times, low memory requirements). But instead of finding a better geometrical discretization of the considered structure, the material parameters are averaged in each sub-volume cell[1]. Indeed, this approach is also implemented in CST Microwave Studio.

---

[1] Since the Perfect Boundary Approximation is a commercially used algorithm, no detailed information about its principles can be found.

Bibliography

[1] E. Hecht, *Optik* (R. Oldenbourg Verlag, 1999).

[2] C. Kittel, *Einführung in die Festkörperphysik* (Oldenbourg, R., Verlag GmbH, 1996), 10th ed.

[3] N. W. Ashcroft and N. D. Mermin, *Solid State Physics* (Oldenbourg, R., Verlag GmbH, 1976), international ed.

[4] W. Koechner, *Solid-State Laser Engineering* (Springer Media, Inc., 2006).

[5] E. Yablonovitch, "Inhibited spontaneous emission in solid-state physics and electronics," Phys. Rev. Lett. **58**, 2059–2062 (1987).

[6] S. John, "Strong localization of photons in certain disordered dielectric superlattices," Phys. Rev. Lett. **58**, 2486–2489 (1987).

[7] K. Busch, G. von Freymann, S. Linden, S. F. Mingaleev, L. Tkeshelashvili, and M. Wegener, "Periodic nanostructures for photonics," Phys. Rep. **444**, 101–202 (2007).

[8] G. von Freymann, A. Ledermann, M. Thiel, I. Staude, S. Essig, K. Busch, and M. Wegener, "Three-dimensional nanostructures for photonics," Adv. Funct. Mater. **20**, 1038–1052 (2010).

[9] V. G. Veselago, "Electrodynamics of substances with simultaneously negative values of ε and μ," Sov. Phys. Uspekhi **10**, 509–514 (1968).

[10] J. B. Pendry, A. J. Holden, D. J. Robbins, and W. J. Stewart, "Magnetism from conductors and enhanced nonlinear phenomena," IEEE Trans. Microw. Theory Tech. **47**, 2075–2084 (1999).

[11] W. N. Hardy and L. A. Whitehead, "Split-ring resonator for use in magnetic-resonance from 200-2000 MHz," Rev. Sci. Instrum. **52**, 213–216 (1981).

[12] D. R. Smith, W. J. Padilla, D. C. Vier, S. C. Nemat-Nasser, and S. Schultz, "Composite medium with simultaneously negative permeability and permittivity," Phys. Rev. Lett. **84**, 4184–4187 (2000).

[13] R. A. Shelby, D. R. Smith, S. C. Nemat-Nasser, and S. Schultz, "Microwave transmission through a two-dimensional, isotropic, left-handed metamaterial," Appl. Phys. Lett. **78**, 489 (2001).

[14] R. A. Shelby, D. R. Smith, and S. Schultz, "Experimental verification of a negative index of refraction," Science **292**, 77–79 (2001).

[15] J. B. Pendry, "Negative refraction makes a perfect lens," Phys. Rev. Lett. **85**, 3966–3969 (2000).

[16] N. Fang, H. Lee, C. Sun, and X. Zhang, "Sub-diffraction-limited optical imaging with a silver superlens," Science **308**, 534–537 (2005).

[17] N. Seddon and T. Bearpark, "Observation of the inverse Doppler effect," Science **302**, 1537–1540 (2003).

[18] J. Lu, T. M. Grzegorczyk, Y. Zhang, J. Pacheco, B.-I. Wu, and J. A. Kong, "Cerenkov radiation in materials with negative permittivity and permeability," Opt. Express **11**, 723–734 (2003).

[19] Z. Duan, B.-I. Wu, and M. Chen, "Review of Cherenkov radiation in double-negative metamaterials," in "Prog. In Electromagn. Res. Symp.", (2009), pp. 65–67.

[20] J. B. Pendry, D. Schurig, and D. R. Smith, "Controlling electromagnetic fields," Science **312**, 1780–1782 (2006).

[21] D. Schurig, J. J. Mock, B. J. Justice, S. A. Cummer, J. B. Pendry, A. F. Starr, and D. R. Smith, "Metamaterial electromagnetic cloak at microwave frequencies," Science **314**, 977–980 (2006).

[22] W. S. Cai, U. K. Chettiar, A. V. Kildishev, and V. M. Shalaev, "Optical cloaking with metamaterials," Nat. Photonics **1**, 224–227 (2007).

[23] T. Ergin, N. Stenger, P. Brenner, J. B. Pendry, and M. Wegener, "Three-dimensional invisibility cloak at optical wavelenghts," Science **328**, 337–339 (2010).

[24] T. J. Yen, W. J. Padilla, N. Fang, D. C. Vier, D. R. Smith, J. B. Pendry, D. N. Basov, and X. Zhang, "Terahertz magnetic response from artificial materials," Science **303**, 1494–1496 (2004).

[25] S. Linden, C. Enkrich, M. Wegener, J. F. Zhou, T. Koschny, and C. M. Soukoulis, "Magnetic response of metamaterials at 100 Terahertz," Science **306**, 1351–1353 (2004).

[26] V. M. Shalaev, "Optical negative-index metamaterials," Nat. Photonics **1**, 41–48 (2007).

[27] C. M. Soukoulis, S. Linden, and M. Wegener, "Negative refractive index at optical wavelengths," Science **315**, 47–49 (2007).

[28] N. Liu, H. C. Guo, L. W. Fu, S. Kaiser, H. Schweizer, and H. Giessen, "Three-dimensional photonic metamaterials at optical frequencies," Nat. Mater. **7**, 31–37 (2008).

[29] N. Liu, L. Fu, S. Kaiser, H. Schweizer, and H. Giessen, "Plasmonic building blocks for magnetic molecules in three-dimensional optical metamaterials," Adv. Mater. **20**, 3859–3865 (2008).

[30] G. Dolling, M. Wegener, and S. Linden, "Realization of a three-functional-layer negative-index photonic metamaterial," Opt. Lett. **32**, 551–553 (2007).

[31] J. Valentine, S. Zhang, T. Zentgraf, E. Ulin-Avila, D. A. Genov, G. Bartal, and X. Zhang, "Three-dimensional optical metamaterial with a negative refractive index," Nature **455**, 376–379 (2008).

[32] M. S. Rill, C. Plet, M. Thiel, I. Staude, G. von Freymann, S. Linden, and M. Wegener, "Photonic metamaterials by direct laser writing and silver chemical vapour deposition," Nat. Mater. **7**, 543–546 (2008).

[33] J. K. Gansel, M. Thiel, M. S. Rill, M. Decker, K. Bade, V. Saile, G. von Freymann, S. Linden, and M. Wegener, "Gold helix photonic metamaterial as broadband circular polarizer," Science **325**, 1513–1515 (2009).

[34] J. D. Jackson, *Klassische Elektrodynamik* (Walter de Gruyter, 1983).

[35] L. D. Landau and E. M. Lifshitz, *Electrodynamics of Continuous Media*, vol. 8 (Butterworth-Heinemann, Oxford, 1984).

[36] C. A. Kyriazidou, H. F. Contopanagos, W. M. Merrill, and N. G. Alexopoulos, "Artificial versus natural crystals: Effective wave impedance of printed photonic bandgap materials," IEEE Trans. Antennas Propag. **48**, 95–106 (2000).

[37] D. R. Smith, D. C. Vier, T. Koschny, and C. M. Soukoulis, "Electromagnetic parameter retrieval from inhomogeneous metamaterials," Phys. Rev. E **71**, 036617 (2005).

[38] C. R. Simovski, I. Kolmakov, and S. A. Tretyakov, "Approaches to the homogenization of periodical metamaterials," in "11th International Conference on Mathematical Methods in Electromagnetic Theory (Kharkiv, Ukraine)," (2006), pp. 41–44.

[39] C. R. Simovski and S. A. Tretyakov, "Local constitutive parameters of metamaterials from an effective-medium perspective," Phys. Rev. B **75**, 195111 (2007).

[40] R. J. Potton, "Reciprocity in optics," Rep. Prog. Phys. **67**, 717–754 (2004).

[41] I. V. Lindell and A. J. Viitanen, "Plane wave propagation in uniaxial bianisotropic medium," Electron. Lett. **29**, 150–152 (1993).

[42] S. He, "Wave propagation through a dielectric-uniaxial bianisotropic interface and the computation of Brewster angles," J. Opt. Soc. Am. A **10**, 2402–2409 (1993).

[43] I. V. Lindell and A. J. Viitanen, "Eigenwaves in the general uniaxial bianisotropic medium with symmetric parameter dyadics," PIER **9**, 1–18 (1994).

[44] S. A. Tretyakov and A. A. Sochava, "Reflection and transmission of plane electromagnetic waves in uniaxial bianisotropic materials," Int. J. Infrared and Millimeter Waves **15**, 829–856 (1994).

[45] A. Pimenov, A. Loidl, K. Gehrke, V. Moshnyaga, and K. Samwer, "Negative refraction observed in a metallic ferromagnet in the Gigahertz frequency range," Phys. Rev. Lett. **98**, 197401 (2007).

[46] K. Zhou, D. Wang, K. Huang, L. Yin, Y. Zhou, and S. Gao, "Characteristics of permittivity and permeability spectra in range of 2–18 GHz microwave frequency for $La_{1-x}Sr_xMn_{1-y}B_yO_3$ (B=Fe, Co, Ni)," Trans. Nonferrous Met. Soc. China **17**, 1294–1299 (2007).

[47] R. Merlin, "Metamaterials and the Landau-Lifshitz permeability argument: Large permittivity begets high-frequency magnetism," PNAS **106**, 1693–1698 (2009).

[48] R. M. Walser, *Introduction to Complex Mediums for Optics and Electromagnetics*, vol. PM123 (SPIE Press, 2003).

[49] D. R. Smith, S. Schultz, P. Markoš, and C. M. Soukoulis, "Determination of effective permittivity and permeability of metamaterials from reflection and transmission coeffi-

cients," Phys. Rev. B **65**, 195104 (2002).

[50] C. L. Giles and W. J. Wild, "Brewster angles for magnetic media," J. Infrared Millimeter Waves **3**, 187–197 (1985).

[51] C. E. Kriegler, M. S. Rill, S. Linden, and M. Wegener, "Bianisotropic photonic metamaterials," IEEE J. Sel. Top. Quantum Electron. **16**, 367–375 (2010).

[52] T. P. Meyrath, T. Zentgraf, and H. Giessen, "Lorentz model for metamaterials: Optical frequency resonance circuits," Phys. Rev. B **75**, 205102 (2007).

[53] M. Husnik, M. W. Klein, N. Feth, M. Konig, J. Niegemann, K. Busch, S. Linden, and M. Wegener, "Absolute extinction cross-section of individual magnetic split-ring resonators," Nat. Photonics **2**, 614–617 (2008).

[54] L. Zhou and S. T. Chui, "Eigenmodes of metallic ring systems: A rigorous approach," Phys. Rev. B **74**, 035419 (2006).

[55] O. Sydoruk, E. Tatartschuk, E. Shamonina, and L. Solymar, "Analytical formulation for the resonant frequency of split rings," J. Appl. Phys. **105**, 014903 (2009).

[56] V. Delgado, O. Sydoruk, E. Tatartschuk, R. Marqués, M. J. Freire, and L. Jelinek, "Analytical circuit model for split ring resonators in the far infrared and optical frequency range," Metamaterials **3**, 57–62 (2009).

[57] T. Fließbach, *Elektrodynamik*, vol. 2 (Spektrum Akademischer Verlag, Heidelberg, 2008), 5th ed.

[58] G. Dolling, C. Enkrich, M. Wegener, J. F. Zhou, and C. M. Soukoulis, "Cut-wire pairs and plate pairs as magnetic atoms for optical metamaterials," Opt. Lett. **30**, 3198–3200 (2005).

[59] G. Dolling, "Design, fabrication, and characterization of double-negative metamaterials for photonics," Ph.D. thesis, Universität Karlsruhe (TH) (2007).

[60] T. A. Klar, A. V. Kildishev, V. P. Drachev, and V. M. Shalaev, "Negative-index metamaterials: Going optical," IEEE J. Sel. Top. Quantum Electron. **12**, 1106–1115 (2006).

[61] P. S. J. Russell, "Interference of integrated Floquet-Bloch waves," Phys. Rev. A **33**, 3232–3242 (1986).

[62] R. Zengerle, "Light propagation in singly and doubly periodic planar waveguides," J. Mod. Opt. **34**, 1589–1617 (1987).

[63] M. Notomi, "Theory of light propagation in strongly modulated photonic crystals: Refractionlike behavior in the vicinity of the photonic band gap," Phys. Rev. B **62**, 10696–10705 (2000).

[64] C. Luo, S. G. Johnson, and J. D. Joannopoulos, "All-angle negative refraction in a three-dimensionally periodic photonic crystal," Appl. Phys. Lett. **81**, 2352–2354 (2002).

[65] E. Cubukcu, K. Aydin, E. Ozbay, S. Foteinopoulou, and C. M. Soukoulis, "Electromagnetic waves: Negative refraction by photonic crystals," Nature **423**, 604–605 (2003).

[66] J. Yao, Z. Liu, Y. Liu, Y. Wang, C. Sun, G. Bartal, A. Stacy, and X. Zhang, "Optical negative refraction in bulk metamaterials," Science **321**, 930 (2008).

[67] M. Wegener, G. Dolling, and S. Linden, "Plasmonics: Backward waves moving forward," Nat. Mater. **6**, 475–476 (2007).

[68] G. Dolling, C. Enkrich, M. Wegener, C. M. Soukoulis, and S. Linden, "Simultaneous negative phase and group velocity of light in a metamaterial," Science **312**, 892–894 (2006).

[69] A. Lakhtakia, "Positive and negative Goos-Hänchen shifts and negative phase-velocity mediums (alias left-handed materials)," Int. J. Electron. Commun. **58**, 229–231 (2004).

[70] T. G. Mackay and A. Lakhtakia, "Negative refraction, negative phase velocity, and counterposition in bianisotropic materials and metamaterials," Phys. Rev. B **79**, 235121 (2009).

[71] H. J. Lezec, J. A. Dionne, and H. A. Atwater, "Negative refraction at visible frequencies," Science **316**, 430–432 (2007).

[72] B. A. Aničin, R. Fazlić, and M. Koprić, "Theoretical evidence for negative Goos-Haenchen shifts," J. Phys. A **11**, 1657–1662 (1978).

[73] P. R. Berman, "Goos-Hänchen shift in negatively refractive media," Phys. Rev. E **66**, 067603 (2002).

[74] D. R. Smith and N. Kroll, "Negative refractive index in left-handed materials," Phys. Rev. Lett. **85**, 2933–2936 (2000).

[75] C. Caloz and T. Itoh, *Electromagnetic Metamaterials: Transmission Line Theory and Microwave Applications* (John Wiley & Sons, Inc., 2006).

[76] P. B. Johnson and R. W. Christy, "Optical constants of the noble metals," Phys. Rev. B **6**, 4370–4379 (1972).

[77] M. A. Ordal, L. L. Long, R. J. Bell, S. E. Bell, R. R. Bell, R. W. Alexander, and C. A. Ward, "Optical properties of the metals Al, Co, Cu, Au, Fe, Pb, Ni, Pd, Pt, Ag, Ti, and W in the infrared and far infrared," Appl. Opt. **22**, 1099–1120 (1983).

[78] J. B. Pendry, A. J. Holden, W. J. Stewart, and I. Youngs, "Extreme low frequency plasmons in metallic mesostructures," Phys. Rev. Lett. **76**, 4773–4776 (1996).

[79] S. Zhang, W. Fan, K. J. Malloy, S. R. J. Brueck, N. C. Panoiu, and R. M. Osgood, "Near-infrared double negative metamaterials," Opt. Express **13**, 4922–4930 (2005).

[80] G. Dolling, M. Wegener, C. M. Soukoulis, and S. Linden, "Negative-index metamaterial at 780 nm wavelength," Opt. Lett. **32**, 53–55 (2007).

[81] M. Wegener and S. Linden, *Tutorials in Metamaterials* (Francis Books, in preparation), chap. Bi-anisotropic and Chiral Metamaterials.

[82] R. Marqués, F. Medina, and R. Rafii-El-Idrissi, "Role of bianisotropy in negative permeability and left-handed metamaterials," Phys. Rev. B **65**, 144440 (2002).

[83] X. Chen, B. I. Wu, J. A. Kong, and T. M. Grzegorczyk, "Retrieval of the effective constitutive parameters of bianisotropic metamaterials," Phys. Rev. E **71**, 046610 (2005).

[84] C. E. Kriegler, "A new route towards 3D metamaterials," Diploma thesis, Universität Karlsruhe (TH) (2008).

[85] G. Dolling, M. Wegener, S. Linden, and C. Hormann, "Photorealistic images of objects in effective negative-index materials," Opt. Express **14**, 1842–1849 (2006).

[86] J. C. Maxwell Garnett, "Colours in metal glasses and in metallic films," Phil. Trans. Royal Soc. London A **3**, 385–420 (1904).

[87] A. Lakhtakia and W. S. Weiglhofer, "Maxwell–Garnett estimates of the effective properties of a general class of discrete random composites," Acta Cryst. A **49**, 266–269 (1993).

[88] D. H. Kwon, D. H.Werner, A. V. Kildishev, and V. Shalaev, "Material parameter retrieval procedure for general bi-isotropic metamaterials and its application to optical chiral negative-index metamaterial design," Opt. Express **16**, 822–829 (2008).

[89] K. Robbie, M. J. Brett, and A. Lakhtakia, "Chiral sculptured thin films," Nature **384**, 616 (1996).

[90] A. Chutinan and S. Noda, "Spiral three-dimensional photonic-band-gap structure," Phys. Rev. B **57**, R2006–R2008 (1998).

[91] S. R. Kennedy, M. J. Brett, O. Toader, and S. John, "Fabrication of tetragonal square spiral photonic crystals," Nano Lett. **2**, 59–62 (2002).

[92] M. Thiel, M. S. Rill, G. von Freymann, and M. Wegener, "Three-dimensional bi-chiral photonic crystals," Adv. Mater. **21**, 4680–4682 (2009).

[93] A. V. Rogacheva, V. A. Fedotov, A. S. Schwanecke, and N. I. Zheludev, "Giant gyrotropy due to electromagnetic-field coupling in a bilayered chiral structure," Phys. Rev. Lett. **97**, 177401 (2006).

[94] M. Decker, M. W. Klein, M. Wegener, and S. Linden, "Circular dichroism of planar chiral magnetic metamaterials," Opt. Lett. **32**, 856–858 (2007).

[95] E. Plum, J. Zhou, J. Dong, V. A. Fedotov, T. Koschny, C. M. Soukoulis, and N. I. Zheludev, "Metamaterial with negative index due to chirality," Phys. Rev. B **79**, 035407 (2009).

[96] D. R. Smith, J. B. Pendry, and M. C. K. Wiltshire, "Metamaterials and negative refractive index," Science **305**, 788–792 (2004).

[97] M. W. Klein, C. Enkrich, M. Wegener, C. M. Soukoulis, and S. Linden, "Single-slit split-ring resonators at optical frequencies: limits of size scaling," Opt. Lett. **31**, 1259–1261 (2006).

[98] U. K. Chettiar, A. V. Kildishev, H.-K. Yuan, W. Cai, S. Xiao, V. P. Drachev, and V. M. Shalaev, "Dual-band negative index metamaterial: Double negative at 813 nm and single negative at 772 nm," Opt. Lett. **32**, 1671–1673 (2007).

[99] K. S. Novoselov, A. K. Geim, S. V. Morozov, D. Jiang, Y. Zhang, S. V. Dubonos, I. V. Grigorieva, and A. A. Firsov, "Electric field effect in atomically thin carbon films," Science **306**, 666–669 (2004).

[100] A. K. Geim and K. S. Novoselov, "The rise of graphene," Nat. Mater. **6**, 183–191 (2007).

[101] B. Partoens and F. M. Peeters, "From graphene to graphite: Electronic structure around the K point," Phys. Rev. B **74**, 075404 (2006).

[102] C. Enkrich, F. Perez-Willard, D. Gerthsen, J. F. Zhou, T. Koschny, C. M. Soukoulis, M. Wegener, and S. Linden, "Focused-ion-beam nanofabrication of near-infrared magnetic metamaterials," Adv. Mater. **17**, 2547–2549 (2005).

[103] E. Tekin, P. J. Smith, and U. S. Schubert, "Inkjet printing as a deposition and patterning tool for polymers and inorganic particles," Soft Matter **4**, 703–713 (2008).

[104] K. Takano, T. Kawabata, C. Hsieh, K. Akiyama, F. Miyamaru, Y. Abe, Y. Tokuda, R. Pan, C. Pan, and M. Hangyo, "Fabrication of terahertz planar metamaterials using a super-fine ink-jet printer," Appl. Phys. Express **3**, 016701 (2010).

[105] Z. Y. Ku, J. Y. Zhang, and S. R. J. Brueck, "Bi-anisotropy of multiple-layer fishnet negative-index metamaterials due to angled sidewalls," Opt. Express **17**, 6782–6789 (2009).

[106] G. Subramania and S. Y. Lin, "Fabrication of three-dimensional photonic crystal with alignment based on electron beam lithography," Appl. Phys. Lett. **85**, 5037–5039 (2004).

[107] E. Plum, V. A. Fedotov, A. S. Schwanecke, and N. I. Zheludev, "Giant optical gyrotropy due to electromagnetic coupling," Appl. Phys. Lett. **90**, 223113 (2007).

[108] M. Decker, M. Ruther, C. Kriegler, J. Zhou, C. Soukoulis, S. Linden, and M. Wegener, "Strong optical activity from twisted-cross photonic metamaterials," Opt. Lett. **34**, 2501–2503 (2009).

[109] C. E. Kriegler, M. S. Rill, M. Thiel, E. Müller, S. Essig, A. Frölich, G. von Freymann, S. Linden, D. Gerthsen, H. Hahn, K. Busch, and M. Wegener, "Transition between corrugated metal films and split-ring-resonator arrays," Appl. Phys. B **96**, 1–7 (2009).

[110] M. S. Rill, C. E. Kriegler, M. Thiel, G. von Freymann, S. Linden, and M. Wegener, "Negative-index bianisotropic photonic metamaterial fabricated by direct laser writing and silver shadow evaporation," Opt. Lett. **34**, 19–21 (2009).

[111] M. S. Rill, C. Plet, M. Thiel, I. Staude, M. Wegener, G. von Freymann, and S. Linden, "Photonic metamaterial structures by 3D direct laser writing," in "Plasmonics and Metamaterials," (2008), talk MTuC3.

[112] M. Rill, C. Plet, M. Thiel, G. von Freymann, S. Linden, and M. Wegener, "Photonic metamaterials by direct laser writing and silver chemical vapor deposition," in "Quantum Electronics and Laser Science Conference (QELS)," (2008), talk QMD3.

[113] M. S. Rill, C. E. Kriegler, M. Thiel, A. Frölich, E. Müller, D. Gerthsen, S. Essig, K. Busch, G. von Freymann, S. Linden, H. Hahn, and M. Wegener, "Photonic metamaterials by direct laser writing," in "International Quantum Electronics Conference (IQEC)," (2009), talk JWE3.

[114] S. Maruo, O. Nakamura, and S. Kawata, "Three-dimensional microfabrication with two-photon-absorbed photopolymerization," Opt. Lett. **22**, 132–134 (1997).

[115] S. Kawata, H. B. Sun, T. Tanaka, and K. Takada, "Finer features for functional microdevices – micromachines can be created with higher resolution using two-photon absorption." Nature **412**, 697–698 (2001).

[116] M. Deubel, G. Von Freymann, M. Wegener, S. Pereira, K. Busch, and C. M. Soukoulis, "Direct laser writing of three-dimensional photonic-crystal templates for telecommunications," Nat. Mater. **3**, 444–447 (2004).

[117] D. J. Ehrlich and J. Melngailis, "Fast room-temperature growth of SiO_2-films by molecular-layer dosing," Appl. Phys. Lett. **58**, 2675–2677 (1991).

[118] M. Ritala, M. Leskelä, E. Nykänen, P. Soininen, and L. Niinistö, "Growth of titanium-dioxide thin-films by atomic layer epitaxy," Thin Solid Films **225**, 288–295 (1993).

[119] S. Haukka, E. L. Lakomaa, and A. Root, "An IR and NMR-study of the chemisorption of $TiCl_4$ on silica," J. Phys. Chem. **97**, 5085–5094 (1993).

[120] M. J. Hampden-Smith and T. T. Kodas, "Chemical vapor deposition of metals: Part 1. An overview of CVD processes," Chem. Vap. Deposition **1**, 8–23 (1995).

[121] E. T. Eisenbraun, A. Klaver, Z. Patel, G. Nuesca, and A. E. Kaloyeros, "Low temperature metalorganic chemical vapor deposition of conformal silver coatings for applications in high aspect ratio structures," J. Vac. Sci. Technol. B **19**, 585–588 (2001).

[122] A. Ishikawa, T. Tanaka, and S. Kawata, "Improvement in the reduction of silver ions in aqueous solution using two-photon sensitive dye," Appl. Phys. Lett. **89**, 113102 (2006).

[123] T. Tanaka, "Plasmonic metamaterials produced by two-photon-induced photoreduction technique," J. Laser Micro/Nanoeng. **3**, 152–156 (2008).

[124] C. N. LaFratta, J. T. Fourkas, T. Baldacchini, and R. A. Farrer, "Multiphoton fabrication," Angew. Chem. Int. Ed. **46**, 6238–6258 (2007).

[125] M. Göppert-Mayer, "Über Elementarakte mit zwei Quantensprüngen," Ann. Phys. **401**, 273–294 (1931).

[126] J. Aarik, A. Aidla, A. A. Kiisler, T. Uustare, and V. Sammelselg, "Effect of crystal structure on optical properties of TiO_2 films grown by atomic layer deposition," Thin Solid Films **305**, 270–273 (1997).

[127] M. Hermatschweiler, A. Ledermann, G. A. Ozin, M. Wegener, and G. von Freymann, "Fabrication of silicon inverse woodpile photonic crystals," Adv. Funct. Mater. **17**, 2273–2277 (2007).

[128] F. Formanek, N. Takeyasu, T. Tanaka, K. Chiyoda, A. Ishikawa, and S. Kawata, "Selective electroless plating to fabricate complex three-dimensional metallic micro/nanostructures," Appl. Phys. Lett. **88**, 083110 (2006).

[129] V. Mizeikis, S. Juodkazis, R. Tarozaitė, J. Juodkazytė, K. Juodkazis, and H. Misawa, "Fabrication and properties of metalo-dielectric photonic crystal structures for infrared spectral region," Opt. Express **15**, 8454–8464 (2007).

[130] A. Tal, Y.-S. Chen, H. E. Williams, R. C. Rumpf, and S. M. Kuebler, "Fabrication and characterization of three-dimensional copper metallodielectric photonic crystals," Opt. Express **15**, 18283–18293 (2007).

[131] N. Takeyasu, T. Tanaka, and S. Kawata, "Fabrication of 3D metal/polymer microstructures by site-selective metal coating," Appl. Phys. A **90**, 205–209 (2008).

[132] T. Aaltonen, "Atomic layer deposition of noble metal thin films," Ph.D. thesis, University of Helsinki, Finnland (2005).

[133] Z. Li and R. G. Gordon, "Thin, continuous, and conformal copper films by reduction of atomic layer deposited copper nitride," Chem. Vap. Deposition **12**, 435–441 (2006).

[134] A. Frölich, "Herstellung dreidimensionaler metallischer Nanostrukturen mit Atomlagenabscheidung von Kupfer," Master's thesis, Universität Karlsruhe (TH) (2009).

[135] J. Li, M. M. Hossain, B. Jia, D. Buso, and M. Gu, "Three-dimensional hybrid photonic crystals merged with localized plasmon resonances," Opt. Express **18**, 4491–4498 (2010).

[136] M. Burresi, D. van Oosten, T. Kampfrath, H. Schoenmaker, R. Heideman, A. Leinse, and L. Kuipers, "Probing the magnetic field of light at optical frequencies," Science **326**, 550–553 (2009).

[137] T. Weiland, "Time domain electromagnetic field computation with finite difference methods," Int. J. Numer. Model **9**, 295–319 (1996).

[138] T. Weiland, M. Timm, and I. Munteanu, "A practical guide to 3-D simulation," IEEE Microw. Mag. **9**, 62–75 (2008).

[139] K. Yee, "Numerical solution of initial boundary value problems involving maxwell's equations in isotropic media," IEEE Trans. Antennas Propag. **14**, 302–307 (1966).

[140] L. Li, "New formulation of the fourier modal method for crossed surface-relief gratings," J. Opt. Soc. Am. A **14**, 2758–2767 (1997).

[141] P. Lalanne and M. P. Jurek, "Computation of the near-field pattern with the coupled-wave method for transverse magnetic polarization," J. Mod. Opt. **45**, 1357–1374 (1998).

[142] D. M. Whittaker and I. S. Culshaw, "Scattering-matrix treatment of patterned multilayer photonic structures," Phys. Rev. B **60**, 2610–2618 (1999).

[143] S. G. Tikhodeev, A. L. Yablonskii, E. A. Muljarov, N. A. Gippius, and T. Ishihara, "Quasiguided modes and optical properties of photonic crystal slabs," Phys. Rev. B **66**, 045102 (2002).

[144] A. Rogers, *Optical and Quantum Electronics 4 – Essentials of Optoelectronics* (Chapman and Hall, 1997), chap. Wiener Khinchin Theorem, p. 144.

[145] D. O. Güney, T. Koschny, M. Kafesaki, and C. A. Soukoulis, "Connected bulk negative index photonic metamaterials," Opt. Lett. **34**, 506–508 (2009).

[146] T. Koschny, L. Zhang, and C. M. Soukoulis, "Isotropic three-dimensional left-handed metamaterials," Phys. Rev. B **71**, 121103 (2005).

[147] J. D. Baena, L. Jelinek, R. Marqués, and J. Zehentner, "Electrically small isotropic three-dimensional magnetic resonators for metamaterial design," Appl. Phys. Lett. **88**, 134108 (2006).

[148] C. Rockstuhl, F. Lederer, C. Etrich, T. Pertsch, and T. Scharf, "Design of an artificial three-dimensional composite metamaterial with magnetic resonances in the visible range of the electromagnetic spectrum," Phys. Rev. Lett. **99**, 017401 (2007).

[149] C. García-Meca, R. Ortuno, R. Salvador, A. Martínez, and J. Martí, "Low-loss single-layer metamaterial with negative index of refraction at visible wavelengths," Opt. Express **15**, 9320–9325 (2007).

[150] A. Andryieuski, R. Malureanu, and A. Lavrinenko, "Nested structures approach in designing an isotropic negative-index material for infrared," J. Eur. Opt. Soc. **4**, 09003 (2009).

[151] J. Shin, J. T. Shen, and S. H. Fan, "Three-dimensional metamaterials with an ultrahigh effective refractive index over a broad bandwidth," Phys. Rev. Lett. **102**, 093903 (2009).

[152] H. Chen, L. Ran, J. Huangfu, X. Zhang, K. Chen, T. M. Grzegorczyk, and J. A. Kong, "Left-handed materials composed of only s-shaped resonators," Phys. Rev. E **70**, 057605 (2004).

[153] X. X. Cheng, H. S. Chen, T. Jiang, L. X. Ran, and J. A. Kong, "Free space measurement of the cross-polarized transmission band of a bianisotropic left-handed metamaterial," Appl. Phys. Lett. **92**, 174103 (2008).

[154] D. O. Güney, T. Koschny, and C. M. Soukoulis, "Intra-connected three-dimensionally isotropic bulk negative index photonic metamaterial," arxiv **1004.0389**, 1–6 (2010).

[155] N. Liu, T. Weiss, M. Mesch, L. Langguth, U. Eigenthaler, M. Hirscher, C. Sönnichsen, and H. Giessen, "Planar metamaterial analogue of electromagnetically induced transparency for plasmonic sensing," Nano Lett. **10**, 1103–1107 (2010).

[156] J. Fischer, G. von Freymann, and M. Wegener, "The materials challenge in diffraction-unlimited direct-laser-writing optical lithography," Adv. Mater. (in press).

[157] R. W. Wood, "On a remarkable case of uneven distribution of light in a diffraction grating," Phil. Mag. **4**, 396–402 (1902).

[158] A. Hessel and A. A. Oliner, "A new theory of Wood's anomalies on optical gratings," Appl. Opt. **4**, 1275–1297 (1965).

[159] R. E. Raab and A. H. Sihvola, "On the existence of linear non-reciprocal bi-isotropic (NRBI) media," J. Phys. A **30**, 1335–1344 (1997).

[160] W. S. Weiglhofer and A. Lakhtakia, "On the non-existence of linear non-reciprocal bi-isotropic (NRBI) media," J. Phys. A **30**, 2597–2600 (1997).

[161] R. E. Raab and A. H. Sihvola, "Reply to the comment on the existence of NRBI media," J. Phys. A **31**, 1111–1112 (1998).

[162] C. M. Krowne, "Marginal nonreciprocity in linear bi-isotropic media," Microw. Opt. Technol. Lett. **18**, 356–359 (1998).

[163] S. A. Tretyakov, A. H. Sihvola, A. A. Sochava, and C. R. Simovski, "Magnetoelectric interactions in bi-anisotropic media," J. Electromagn. Waves Appl. **12**, 481–497 (1998).

[164] E. O. Kamenetskii, "Nonreciprocal microwave bianisotropic materials: Reciprocity theorem and network reciprocity," IEEE Trans. Antennas Propag. **49**, 361–366 (2001).

[165] S. Tretyakov, A. Sihvola, and B. Jancewicz, "Onsager-Casimir principle and the constitutive relations of bi-anisotropic media," J. Electromagn. Waves Appl. **16**, 573–587

(2002).

[166] A. Taflove, "Application of the finite-difference time-domain method to sinusoidal steady-state electromagnetic-penetration problems," IEEE Trans. Electromagn. Compat. **EMC-22**, 191–202 (1980).

[167] J. S. Hesthaven and T. Warburton, "Nodal high-order methods on unstructured grids: I. Time-domain solution of Maxwell's equations," J. Comp. Phys. **181**, 186–221 (2002).

[168] K. Stannigel, M. König, J. Niegemann, and K. Busch, "Discontinuous Galerkin time-domain computations of metallic nanostructures," Opt. Express **17**, 14934–14947 (2009).

[169] K. Busch, J. Niegemann, M. Pototschnig, and L. Tkeshelashvili, "A Krylov-subspace based solver for the linear and nonlinear Maxwell equations," Phys. Status Solidi B **244**, 3479–3496 (2007).

[170] R. Holland, "Finite-difference solution of Maxwell's equations in generalized nonorthogonal coordinates," IEEE Trans. Nucl. Sciences **30**, 4589–4591 (1983).

[171] R. Schuhmann and T. Weiland, "Stability of the FDTD algorithm on nonorthogonal grids related to the spatial interpolation scheme," IEEE Trans. Magn. **34**, 2751–2754 (1998).

[172] J. Niegemann, M. König, K. Stannigel, and K. Busch, "Higher-order time-domain methods for the analysis of nano-photonic systems," Photonics Nanostruct. Fundam. Appl. **7**, 2–11 (2009).

[173] B. Krietenstein, R. Schuhmann, P. Thoma, and T. Weiland, "The Perfect Boundary Approximation technique facing the challenge of high precision field computation," in "XIX International Linear Accelerator Conference (LINAC 98)," (1998), pp. 860–862.

Acknowledgments

First of all, I would like to thank my Ph.D. supervisor Prof. Dr. Martin Wegener for giving me the opportunity to work on such an interesting topic. The research conditions in his group are to the best advantage. This Thesis has largely benefited from his continuous interest, numerous physical and technical discussions as well as his profound physical background.

Moreover, I would like to thank Prof. Dr. Kurt Busch for kindly agreeing to co-referee this Thesis. For sure, it is very helpful for experimentalists to have access to experts in numerics sitting in the reaching area of the hands. The joint publications have highly benefited from his and his group members' profound theoretical knowledge in the field of metamaterials.

I would like to express my gratitude and appreciation to the following co-investigators:

- Prof. Dr. Stefan Linden, my room mate during the last years, who highly contributed to many physical discussions. He gave me many important hints and advices concerning theoretical and experimental questions.
- I am grateful to my former diploma students Christine E. Kriegler and Andreas Frölich who contributed so much to the results of my Thesis. I really appreciate their patience and assiduity. They are now Ph.D. students by themselves and I am pretty sure that their theses will be incredibly good.
- Special thanks to my team colleagues Sabine Essig, Justyna K. Gansel, Erich Müller, Isabelle Staude, and Dr. Michael Thiel who have been directly involved in my scientific projects.
- I am indebted to the proof readers of my Thesis, i.e. (in alphabetical order), Dr. Georg von Freymann, Justyna K. Gansel, Christian Hopfensitz, Martin Husnik, Christine E. Kriegler, Prof. Dr. Stefan Linden, and Dr. Michael Thiel. Their supporting comments and numerous hints really helped to improve the readability and quality of this manuscript.

Next, I would like to thank all the other and former members of the Wegener group for their support, the amicable atmosphere, and for sharing many beautiful moments with me, in and outside the office.

Further cordial thanks go also to the different facilities and people of the Institut für Angewandte Physik, i.e. (in alphabetical order), Patrice Brenner, Werner Gilde, Renate Helfen, Gisela Habitzreither, Jacques Hawecker, Heinz Hoffmann, Frank Landhäußer, Helmut Lay, Heinz Leonhard, Werner Wagner, Christa Weißenburger, and Johann Westhauser. Especially, I would like to express my gratitude to our former technician and friend Thorsten Kuhn. We all miss Thorsten very much!

I also would like to thank the DFG-Center for Functional Nanostructures (CFN), the Karlsruhe School of Optics and Photonics (KSOP), and the PHOME project for their financial support during my Thesis.

My last words are devoted in gratitude towards my parents and nearest friends who helped to keep the head up during hard times.

I want morebooks!

Buy your books fast and straightforward online - at one of world's fastest growing online book stores! Environmentally sound due to Print-on-Demand technologies.

Buy your books online at
www.morebooks.shop

Kaufen Sie Ihre Bücher schnell und unkompliziert online – auf einer der am schnellsten wachsenden Buchhandelsplattformen weltweit! Dank Print-On-Demand umwelt- und ressourcenschonend produziert.

Bücher schneller online kaufen
www.morebooks.shop

KS OmniScriptum Publishing
Brivibas gatve 197
LV-1039 Riga, Latvia
Telefax: +371 686 204 55

info@omniscriptum.com
www.omniscriptum.com

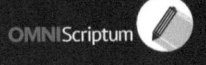

Printed by Books on Demand GmbH, Norderstedt / Germany